Time Bank as a Complementary Economic System:

Emerging Research and Opportunities

Lukas Valek
University of Hradec Kralove, Czech Republic

Vladimir Bures
University of Hradec Kralove, Czech Republic

A volume in the Advances in Finance, Accounting, and Economics (AFAE) Book Series

IGI Global
DISSEMINATOR OF KNOWLEDGE

Published in the United States of America by
IGI Global
Business Science Reference (an imprint of IGI Global)
701 E. Chocolate Avenue
Hershey PA, USA 17033
Tel: 717-533-8845
Fax: 717-533-8661
E-mail: cust@igi-global.com
Web site: http://www.igi-global.com

Library of Congress Cataloging-in-Publication Data

Names: Valek, Lukas, 1982- author. | Bures, Vladimir, 1977- author.
Title: Time bank as a complementary economic system : emerging research and opportunities / by Lukas Valek and Vladimir Bures.
Description: Hershey, PA : Business Science Reference, [2019]
Identifiers: LCCN 2018011078| ISBN 9781522569749 (hardcover) | ISBN 9781522569756 (ebook)
Subjects: LCSH: Time banking. | Exchange. | Currency question. | System analysis.
Classification: LCC HG255 .V35 2019 | DDC 332/.5--dc23 LC record available at https://lccn.loc.gov/2018011078

This book is published in the IGI Global book series Advances in Finance, Accounting, and Economics (AFAE) (ISSN: 2327-5677; eISSN: 2327-5685)

British Cataloguing in Publication Data
A Cataloguing in Publication record for this book is available from the British Library.

Advances in Finance, Accounting, and Economics (AFAE) Book Series

ISSN:2327-5677
EISSN:2327-5685

MISSION

In our changing economic and business environment, it is important to consider the financial changes occurring internationally as well as within individual organizations and business environments. Understanding these changes as well as the factors that influence them is crucial in preparing for our financial future and ensuring economic sustainability and growth.

The **Advances in Finance, Accounting, and Economics (AFAE)** book series aims to publish comprehensive and informative titles in all areas of economics and economic theory, finance, and accounting to assist in advancing the available knowledge and providing for further research development in these dynamic fields.

COVERAGE

- Field Research
- Economic Theory
- Microeconomics
- Applied Accounting
- Macroeconomics
- Bankruptcy
- Evidence-Based Studies
- Economics of Risks, Uncertainty, Ambiguity, and Insurance
- Stock Market
- Economics of Migration and Spatial Mobility

IGI Global is currently accepting manuscripts for publication within this series. To submit a proposal for a volume in this series, please contact our Acquisition Editors at Acquisitions@igi-global.com or visit: http://www.igi-global.com/publish/.

The Advances in Finance, Accounting, and Economics (AFAE) Book Series (ISSN 2327-5677) is published by IGI Global, 701 E. Chocolate Avenue, Hershey, PA 17033-1240, USA, www.igi-global.com. This series is composed of titles available for purchase individually; each title is edited to be contextually exclusive from any other title within the series. For pricing and ordering information please visit http://www.igi-global.com/book-series/advances-finance-accounting-economics/73685. Postmaster: Send all address changes to above address.

Titles in this Series

For a list of additional titles in this series, please visit:
https://www.igi-global.com/book-series/advances-finance-accounting-economics/73685

Microfinance and Its Impact on Entrepreneurial Development, Sustainability, and Inclusive ...
Ramesh Chandra Das (Katwa College, India)
Business Science Reference • ©2018 • 385pp • H/C (ISBN: 9781522552130) • US $245.00

Maintaining Sustainable Accounting Systems in Small Business
Luísa Cagica Carvalho (Open University, Portugal & University of Évora, Portugal) and Elisa Truant (University of Turin, Italy)
Business Science Reference • ©2018 • 387pp • H/C (ISBN: 9781522552673) • US $215.00

Agricultural Finance and Opportunities for Investment and Expansion
Augustine Odinakachukwu Ejiogu (Imo State University, Nigeria)
Business Science Reference • ©2018 • 236pp • H/C (ISBN: 9781522530596) • US $195.00

Regaining Global Stability After the Financial Crisis
Bruno Sergi (Harvard University, USA) Filip Fidanoski (University of New South Wales, Australia) Magdalena Ziolo (University of Szczecin, Poland) and Vladimir Naumovski (University American College Skopje, Macedonia)
Business Science Reference • ©2018 • 383pp • H/C (ISBN: 9781522540267) • US $215.00

Employment Protection Legislation in Emerging Economies
Samir Amine (Universite du Quebec en Outaouais, Canada)
Business Science Reference • ©2018 • 301pp • H/C (ISBN: 9781522541349) • US $195.00

Fractal Approaches for Modeling Financial Assets and Predicting Crises
Inna Nekrasova (Southern Federal University, Russia) Oxana Karnaukhova (Southern Federal University, Russia) and Bryan Christiansen (PryMarke, LLC, USA)
Business Science Reference • ©2018 • 306pp • H/C (ISBN: 9781522537670) • US $215.00

For an entire list of titles in this series, please visit:
https://www.igi-global.com/book-series/advances-finance-accounting-economics/73685

IGI Global
DISSEMINATOR OF KNOWLEDGE

701 East Chocolate Avenue, Hershey, PA 17033, USA
Tel: 717-533-8845 x100 • Fax: 717-533-8661
E-Mail: cust@igi-global.com • www.igi-global.com

Table of Contents

Preface

One important way to study world around us in all its complexity is to look at it as a system, thus the methodologies of systems thinking should prove to be a logical approach to a systemic analysis. The wider system of the space around humans is composed of interacting sub-systems which in turn are composed of other sub-systems. Some systems have predictable causality and can be decomposed, analyzed and predicted quite easily. Some include unpredictable elements such as human beings. The study of these so-called "soft systems" ranks among the most difficult tasks for systems engineering, as soft systems are ever-changing, thus unpredictable and hard to quantify and qualify by experiment. This book focuses on a systemic study of one such soft system called the Time Bank. The Time Bank is an exchange concept based on a single economic template, yet has countless uses in various fields of human activity. The aim of this book is to show this universality by research, systems analysis and synthesis as well as various other methods. Although researches have been conducted in the field of Time Banking before, this book attempts to be unique in its scope, methodology and overall approach, and thus seeks to offer a much more complete systemic overview of the issue than any previous researches.

Contemporary research in field of Time Bank has generally been unstructured and takes a retrospective point of view. Available studies tend to analyze the Time Bank by describing and analyzing already extant approaches, the pros and cons thereof, along with impacts on target groups, the environment and society in general, etc. rather than taking a systemic point of view, i.e. conceiving a particular Time Bank as a system and applying it. In practice, approaches to the Time Bank commonly taken until now have shown that there can be as many points of view as there are researchers. At first glance, this retrospective and particularized approach would seem to make sense, as one of most prominent features of the Time Bank is its universality and adaptability. In my opinion, however, books and articles produced by this research lack a solid theoretical basis in terms of definition and structure, i.e. with this approach the Time Bank has been studied merely on an ex-post basis rather than on a conceptual level.

As with any other economic system, a Time Bank is composed of certain key elements, which are generally defined by the people who are organized or, perhaps better said, organize themselves into a particular structure. The Time Bank belongs to the family of complementary economic systems (CES) and are defined by a set of main core values, yet each must be greatly adapted to local conditions, which makes categorizing them complicated, especially as a structured system. In other words, Time Banking lacks both methodological roots and a clear definition as a system. CES, including Time Banks, are supplements to mainstream economic approaches, and they fill the empty spaces where society needs them to do so. The relation of Time Banks to the systemic sciences remains uncharted territory, with connections among various Time Banking phenomena having been undertaken in only few theoretical works; utilizing a TB has generally been considered merely as one of several possible direct, practical approaches to solving societal problems, rather than as an area for theoretical or speculative research (Gregory, 2015).

This book assumes a different perspective. Rather than studying the Time Bank concept in mechanistic way as a chain of causality, it looks for answers in its very systemic structure. As indicated, the usual approach to research is to study one particular existing Time Bank and its impact. The approach of this book begins with background research regarding the basic concept itself – its roots, types, previous basic researches conducted, as well as the TB's current position in science – to formulate a description using a extensive collection of all the latest data available from a number of disciplines. In the next stages of the research, the available approaches are classified, systemic elements determined, and systems analysis conducted from which overall conclusions are drawn regarding broader possible conceptions of TB. This global approach seeks to offer a previously unseen interpretation of the issues involved as well as to produce several results which can be applied in as many spheres of life and the sciences as the Time Bank can be used. And because Time Banks can in fact be used universally – which will be demonstrated in this book – the applications follow this universality.

As mentioned above, the book starts with providing the contemporary context regarding the Time Bank and describing the most recent research methodologies and results. The "state-of-the-art" chapter contains a short contextual background as well as ties to other disciplines, finally focusing on systems engineering along with knowledge and information management.

Nevertheless, as this approach and perspective is quite new, it is necessary to describe the whole context and the roots of Time Banking from its very beginning. In clarifying its origins in this manner, its potential applications and relations to scientific activity can be demonstrated and justified.

The next chapter focuses on methodology, describing the huge amount and varieties of strategies used throughout the research due to the complexity of the subject. In order to tackle Time Bank as a soft system, a specifically relevant methodology of analysis had to be chosen, i.e. one using Peter Checkland's Soft System Methodology (SSM). SSM is used to outline the succeeding steps in the present analysis of the Time Bank approach, which are then used as a guide for the next part of this book. SSM encompasses a comprehensive logical approach to the analysis of the essence of a soft system. Following chapter then offers some insights into the new universal Time Bank system and its features. Moreover, its usage in the selected disciplines of business administration - the management of information and knowledge within the system, and possible uses of information technologies - is outlined. The book is concluded by underlining the main benefits and limitations of the research, the key tangible results of this book and proposals of steps for future inquiries.

The core abbreviations and notations used throughout this book are as follows:

- A TB stands for one particular Time Bank facility
- The TB refers to the institution in general
- TB (no article) indicates the abstract concept of Time Banking
- TBs represents Time Banks (plural).

The medium of exchange in Time Banks is generally called Time Credits, thus the abbreviation TC will be used. The word "member" or "user" indicates a person who is actively involved in Time Bank, whether the institution consists of a physical or legal body. The term "coordinator" is widely used to refer to a person who organizes and manages a Time Bank. In some sources a coordinator is referred to as a "broker" (Simon, 2004).

This book provides the answer to fundamental question associated with TB research: Is there one general approach available for the establishment of the Time Banking concept/system? The main aim is to describe TB as a soft system, identify its elements, analyze them using variety of methods in an attempt to make clear the complexities of the situation in order to determine a universal interpretation to the TB concept which can then be subsequently applied to various areas of human activities.

Objectives:

- To provide relevant holistic background information about TB which would allow a progression toward a systemic analysis
- To find an appropriate methodology by which to analyze TB as a soft system
- To identify all the elements of a TB system
- To design a universal TB concept based on core elements
- To then identify the main characteristics of this universal system
- To identify the main knowledge types and flows in TB related to the core elements
- To identify the main information flows in TB related to the core elements
- To locate contexts within the area of information technologies related to the above findings

REFERENCES

Gregory, L. (2015). *Trading time: Can exchange lead to social change?* Policy Press at the University of Bristol.

Simon, M. (2004). *On becoming a time broker.* Gloucester, UK: Active Communities.

Acknowledgment

The financial support of the Faculty of Informatics and Management, University of Hradec Králové (FIM UHK) Specific Research Project "Socio-Economic Models and Autonomous Systems" is gratefully acknowledged. Authors would like to thank Tomáš Nacházel for this helpfulness and assistance.

Chapter 1
Theoretical Starting Points

ABSTRACT

Every research topic needs to be put in the context of existing theoretical background knowledge. To provide a full understanding of this book, this chapter provides brief insight into selected elementary background theories. Special attention is devoted to the concept of money, which is closely related to the critique of economic development in the 20th and the beginning of the 21st century. Rather than repeating basic economic theory, rhetoric used in the area of complementary economies from which the TB sprouts is introduced. After description of selected complementary economic systems and their relationship to TB, core values and concept of co-production, which are the rationale of any TB definitions, are uncovered. The context description should be comprehensive. Therefore, the additional insight into the issue of law and taxation, TB's societal function, and the role of information technologies is provided.

INTRODUCTION

In order to present a CES version of TB in systemic way, extensive research was necessary. The following pages will introduce connections to most important branches of the social sciences for this book in order to "root" it in the frame of its origins. This would take us from the economic basis of the idea of TB through connections to other social sciences, finishing with a systems approach. With its reach across various sciences the information in the next pages might seem complex, but this analytic approach is necessary

DOI: 10.4018/978-1-5225-6974-9.ch001

in order to provide a background for the analysis and following sections within theoretical specializations. Key relationships are thus unveiled in to provide a wider understanding of the context of the TB concept. Therefore, we will start with theories about money, progressing towards ideas concerning complementary economics supporting this idea using anthropological constructs, ultimately uncovering the values and ideas behind TB.

The environments of complementary economic systems is currently a primary focus of researchers from around the world, with time currency along with Time Banks emerging as key concepts (CCRC, 2016). Despite of the wide spread of the ideas of CES along with the extensive research background, those who decide to implement the idea are generally motivated by its immediate application benefits (Valek, 2016b). This leads to a situation in which new CES are located in research outside wider frameworks, and are then studied on ex-post basis. This could and does in fact lead to quite erratic approaches in terms of implementation, i.e. applications often occur without taking into consideration larger contexts and more extensive systemic ramifications. Certain researches have tried to follow up on these meager bases, but they have mostly remained limited to foundations and local economic contexts along with general possible positive or negative impacts on the broader society. To my knowledge, no comprehensive research exists which employs system sciences. This deficiency has hindered any attempts to grasp the issue of TB from a systemic point of view in terms of its roots, an investigation which would allow broader areas of application.

BACKGROUND

To keep this book logical and consistent it was necessary to start from basics. This leads us from fundamental discourses about money and non-mainstream economic approaches, to complementary economies, and, proceeding from these premises, finally to Time Banks. This progression from basics of money and argumentation related ton complementary economic systems is basic root for understanding role and place of the TB in socio-economic context. Only after that the TB can be described more in detail. The socio-economic context will follow outlining and framing the TB. Some elemental economic theories directly related to contemporary market economy such as Adam Smiths definition of valuation are not mentioned. This is on purpose, because our aim is not to summarize theoretical economic background and defend and argue place of the TB on the face of Earth, rather we bring new points of view. We are aware that reader might expect the issue of TB to be explained in first chapters, but as it will be seen on following pages, it is not entirely possible. In fact, definition of the TB is one of objectives of this book and final answer is to be found only after all analyses in the discussion chapter towards the end of the book. Therefore, we kindly ask readers for patience considering the fact that consecutive order of chapters logically connects facts from various fields finally leading to answers on aim and objectives of this book.

Money

We can find traces of the use of money in various forms throughout the history of humankind. Money is related to economy, with the oldest form of economy believed to be the gift economy (Lietaer, 2001). Money in its origins and earlier forms served to fulfil mysteries of the sacred. It was more a form of payment than exchange, as it was used mainly to pay the initiation fees of religious groups, marriage dues, compensation for fighting or as a tribute to chiefs (Lietaer, 2001). Only much later did the roles of money come to be divided into four areas of use: medium of exchange, measure of value, standard of value, and store of value (Jevons, 1875). As the first type of medium, it allows the exchange of goods with different values. As a measure of value, money sets a particular worth for goods. As a standard of value, it presents the possibility to be lent and repaid at the same worth; as a universal medium, it can store value. For a long time putative commodity currency was used, i.e. a medium which was backed by some commodity (usually a metal, most often gold) and was transferable in this form. Nowadays commodity backing is no longer in place, and what is used is fiat money, a currency not bound to any commodity and is regulated only by laws. In other words, the worth of fiat money depends on the belief of people in it as a currency, as well as the fact that it is widely used for taxation. To a layman it might seem strange that this money exists "out of nothing," standing only on the trust of people, its users. This phenomenon will be further explored later in this text.

In connection to these concepts, arguments have also naturally emerged as to whether it would be wise to return to commodity money, as for example rates of inflation seem to be much less controllable when fiat currency is used (White, 2015). Another feature of money which is often not recognized is that it is actually itself an information system. Compared to ancient times, when bookkeeping was first initiated, the immense amount of transactions that take place daily on a global scale is extraordinarily difficult to describe or even to imagine in real terms. From the point of view of information management, it is interesting to note that money is probably the most pervasive information system, i.e. as flows of real and virtual currencies through billions of transactions in all strata of society (Lietaer, 2001).

Non-Mainstream Economic Approaches in the Digital Age: Complementary Economies

The form of currency generally utilized today is that of the fiat currency. Fiat money is in fact created "out of nothing" and the name itself refers to its enigmatic existence (from Latin Fiat lux in the Biblical book of Genesis, meaning "Let there be light," which can be restated as the decree "Let there be money.") (Lietaer, 2001). Critiques of the fiat currency issued by private institutions through the creation of debt have gone as far as to compare it to arsenic, which although it is deadly poison was used to alleviate short-term symptoms in medicine when there was no better solution (Lietaer et al., 2012). CES are supportive means which provide contextual solutions to problems which come forth the rest of the economy by the creation of fiat money. An early critique of fiat currency was put forth in the 1930s by Irving Fisher, who proposed the so-called "Chicago Plan," the aim of which was to gradually transfer the responsibility of issuing of fiat currency to governments by forcing commercial banks to back deposits by credits from a government and later lowering the amount of money in circulation by lowering the banks' capital deposits. In the working paper "The Chicago Plan Revisited," authors Jaromir Benes and Michael Kumhof re-evaluate the idea, finding that it may carry many benefits (Benes and Kumhof, 2012), but in the end it would be merely the transfer of the authority of the "creation" of money to a single institutional body instead of allowing multiple authorities to control it. Nevertheless, a higher order might be reached by real control over the issuing of fiat money through the prevention of the violent reactions of commercial banks to market events, which is what the Chicago Plan, never instituted, would have offered (Valek, 2018).

Related to the previous statements, another observation that can be added is that fiat money has also now become a commodity (Caroll and Belloti, 2015), although it is difficult to imagine trading with an entity which does not really exist in material form. Tied to concepts like these, the focus of market economies on economic growth has been long institutionalized, despite the fact that endless unlimited growth has been demonstrated to not be possible (Seyfang and Longhurst, 2012, Holmgren, 2011). No subsystem – as is an economic system within the overall system of a society – which is bound by limited material resources – as is the situation on this planet – can grow indefinitely. So it seems we stand on the break of an apparent paradigm change in evaluating economic systems, a transformation in which CES will undoubtedly assume a more important role. This break seems to be emerging primarily at the grassroots level of economies (Seyfang and Longhurst, 2012), i.e. "from the bottom up." It seems the change is in the hands of ordinary people making small changes in their immediate vicinity, e.g. in a regional context (Valek, 2018).

As Nobel-prize winner Wassily Leontieff once said: "The role of humans as the most important factor of production is bound to diminish in the same way that the role of horses in agricultural production was first diminished and then eliminated by introduction of tractors" (Rifkin, 1995). In this analogy, horses may be allowed to die out peacefully, but what we are going to do with humans? (Lietaer, 2001). Humans themselves, as inventors of the "one-way" economic system valorizing never-ending growth, should be able to invent other more sustainable and more humane ways of economy in order to adapt to future societal and technological developments. With the constant development of information technologies in the putative "knowledge society" we might also hope to aim toward a period of transformation, one which Lietaer (Lietaer, 2001) has referred to as a "Wisdom Age," with a concurrent goal of rational growth targets in terms of future economic development (Valek, 2018).

These previous lines might seem utopic, but they are put forward here as only a concentrated description of the kind of rhetoric that many of the founders of complementary economic systems use, and from this somewhat moralistic critique the whole scope of what has been referred to as the non-mainstream economy emerges. History has proven that alternatives to the monetary systems currently in use usually do not work, whereas complements to this system actually possess the power and potential to utilize what is "left behind" by the market economy (Seyfang, 1999, Kennedy et al., 2012). Several examples have emerged, starting with complementary currencies, through exchange systems, unconditional minimal revenue, Time Banks, gift economies and other structures. Many of these options were first mentioned in the literature decades (or much longer) ago, but the fact that they not only persist but have been shown to thrive in many parts of the world shows their feasibility. This book attempts to demonstrate that perhaps the right time has come to introduce them on a larger scale, as they can be – or in fact already have been – refined in suitable working conditions and environments. The following subchapters present some of these complementary structures (Valek, 2018).

Complementary Currencies

The difference between the words "complementary" and completely "alternative" should be quite clear from the previous paragraph. An alternative structure would indicate one that completely replaces the present system with a new one, and a complement would serve to provide support in specific places and particular situations in which the current economy is failing (Lietaer and Dunne, 2013, Schroeder et al., 2011). Complements are easier to integrate into the economy, and by their influence they shift the economic paradigm slowly (Holmgren, 2011). An alternative would have to offer a completely new set of solutions for the whole system; therefore, it is unlikely that this type of "revolutionary" approach would really be operable. It is more likely that an alternative system would be introduced by the slow influence of the complements (Valek, 2018). The Complementary Currency Resource Center, for example, is dedicated to research into complementary currencies (CCRC, 2016).

Most Common Complementary Economies

Local Currencies: These structures represent regionally developed and used currencies of which the main aim is to facilitate exchange within a certain geographically defined region. They have several objectives (Kennedy et al., 2012, Valek, 2018):

Creating local jobs:

- Stem the drain of purchasing power from the region.
- To open up new avenues to enable local government to fulfil its designated tasks

Helping to develop:

- Economic stimulation instead of stagnation
- Stabilization or growth of the population

- Increase in regional purchasing power and, by this, improvement in municipal finances
- Positive identification with the region
- Improvements to the infrastructure and increased local autonomy
- A greater sense amongst the population, a sense of participation
- Preservation and development of employment opportunities and incentives for firms to remain in the region

Local currencies stand upon the trust of their users, which is common feature for all of the complementary currencies. As mentioned above, the main focus is on bringing back activity, prosperity and jobs to a certain region, either urban or rural.

- **LETS (Local Economy Trading System):** LETS represents one of the first working concepts of a trading system, i.e. an economy which grows from the needs of a community. It resembles a local currency system, as it uses one as a medium, but it sprouts from a need to maintain trading within communities in times of crises (Lietaer, 2001). Standard money might be used also in these transactions, but most of exchange is done for the LETS currency (usually a work/service). In this sense, citizens in a particular region can keep trading and working even when official currency is scarce. Scarcity does influence LETS money, as it is generated in the moment when a demand is created, more precisely when a trade is agreed upon. Moreover, currency in a closed system of LETS appears in a strictly reciprocal operation between two parties when at a given time a certain number of units is subtracted on one side and added to on the other side (Martignoni, 2015, Valek, 2018).
- **Business Exchange:** This regional community currency allows small businesses and individuals to continue to trade and produce even when they would not have enough financial power in official currency. A business exchange provides credits, in turn allowing trades which would not normally take place. In addition, a safe environment is created which fosters commerce between and among established business partners. In this sense, it plays a similar supportive role as does a TB in business. One good examples is Swiss WIR, the oldest continuous complementary currency system in the Western world, in existence since 1934 (Lietaer, 2001). The WIR exchange includes both businesses and individuals in a system which in 2010 had an annual turnover rate of 1.6 billion WIR francs (1 WIR franc = 1 Swiss franc).

Other examples of a business exchange have emerged in the RES in Belgium with its yearly turnover of the equivalent of 35 million Euro, and the Business Exchange Scotland, along with others (Kennedy et al., 2012, Valek, 2018).

- **Electronic Currencies:** In the 21st century, many services which lie within the domain of large companies, amongst these banking facilities, have come much closer to ordinary people. Moreover, they can be "grown" from grassroots (Seyfang and Longhurst, 2012) instead of installed from top. The appearance of open source systems has by co-production (introduced below) allowed small enterprises or even individuals to develop their own software, including operating systems by co-production, a fact which has also had a huge impact in the field of telecommunications. Individuals can for the first time even legally circumvent banks in domains which had previously remained untouchable. As one example for all, in the money transfer area the project Transferwise (Transferwise, 2016) allows the complete avoidance of banks and their fees by matching sending and receiving transactions in different countries and currencies, so that the actual payment never leaves the country of origin (Valek, 2018).

Probably the biggest impact in electronic currencies in past few years has been made by the Bitcoin. This public electronic currency founded in 2009 (Bitcoin.cz, 2016) is based on the open source approach. The main critique of the founders of Bitcoin against ordinary money electronic transactions was focused on the necessity within large financial institutions of an intermediary or "middle man" in electronic transactions, with the accompanying cost increases, especially transaction fees, that limit and in fact often completely rule out small casual transactions. Also, the necessity of officially verifying the identities and financial status of the parties involved and with this the requirements to submit personal information slow down and complicate transactions. The proposed approach was an electronic payment system based on cryptographic proof instead of "trust" or identity verification, allowing any two willing parties to transact directly with each other without the need for a trusted third party (Nakamoto, 2009) third party. The verification factor and need for the tracking of exchanges is also inherent for TB, as verification requires memory, while paper money leaves no traceable trail. The Time Dollar, or time credit, is counted as an electronic currency, a fact which builds trust, as the present is shaped by the future (Cahn, 2000). Another feature of the Bitcoin is that the final resource of money is limited, but the currency can be divided into fractions. By this limitation, it comes much closer to a commodity-based currency. The payment cryptographic engine uses so-called "mining" to create new coins.

However, the Bitcoin system is not without its critics. Firstly, it removes the power over money from banks and places it into uncertain hands of volunteers who run the mining engine on their computers, the combined power of which exceeds the world's 500 fastest supercomputers put together (Economist, 2013), thus the security of transactions may be called into question. At any rate, the success of Bitcoin has paved the way for other similar currencies, which mimic it or have tried to change some features of it, e.g. Litecoin, which offers faster transactions, but keeps the limited coin supply feature. Peercoin, of which there is unlimited supply, but comes with implicit 1% inflation. Anoncoin and Zerocoin both aim for complete anonymity of the transactions. In conclusion, although Bitcoin has its own disadvantages, it (and its popularity) have proven a point, that this might be the way of imagining and putting into use new kind of money. The decentralization involved moves the power over currency from banks to individuals. In the same way as, (the eventually found to be illegal) Napster paved the way for services like iTunes and Spotify (Economist, 2013), the road has paved by Bitcoin for new currency independence.

Even though complementary economic systems, including electronic and crypto currencies, have often been met with skepticism, even ridicule, recent developments indicate that in fact we are most likely are at a paradigm break point, as suggested above by Lietaer (Lietaer, 2001). As Bitcoin seemed to spike in value in 2017, Ethereum, another crypto currency, is progressing as a new way of arranging contracts in broader business areas. Of course, these developments might just be passing trends. Nevertheless, the shift in feeling toward this new way of trade is illustrated by foundation of Enterprise Ethereum Alliance (EEA), the membership base of which consists of more and more important enterprises such as Accenture, Cisco, Microsoft, Intel, National Bank of Canada, Santander, MasterCard and many more (EEA, 2018a). The presence of companies of this level and their trust in Ethereum suggests that Ethereum is not merely a fly-by-night consortium of "heretics" (Cahn, 2000, Lietaer et al., 2012). Promoters of complementary currencies were not so long ago called mere utopists, but this medium has become a plausible and even viable development regarding financial services in the digital age. The composition of workgroups within EEA also indicate certain directions in which this development might proceed, e.g. healthcare, communications, banking, insurance, the legal industry, supply chains, energy, etc. (EEA, 2018b).

- **Sharing-Based Economies:** Several systems of exchange deserve mention here. These sharing systems actually decrease costs for users by virtue of the medium of exchange being shared, thus limiting consumption. Typical examples are car sharing, services like Airbnb and Couchsurfing (through which accommodation is shared), tools sharing (along with other various items), etc. through which people use certain items only occasionally and can borrow them from the community for a limited period of time (Valek, 2018).
- **Gift Economy:** This would seem to be the oldest kind of the economy, according to anthropologist Stanley Ambrose of the University of Illinois. Gift giving also helps to strengthen social ties, especially in tough times, and it is particularly important in a demanding and unpredictable environment (Lietaer, 2001). In the Czech Republic, one gift economy-based platform has emerged called Hearth.net (Hearth.net), a web platform which creates a community of people who are willing to give. Those who have something to offer (which could be practically anything) give it unconditionally, and those who need ask from the community that it should be given to them (Valek, 2018).

TIME BANKS

This chapter explains what a TB actually is, i.e. how the concept is currently in use throughout the world. This explanation begins with a description of how the system works, and progresses towards associated values and kinds of TBs. TB is believed to be unique among community currency models (Clement et al., 2017). Deeper historical connotations are elucidated with links to theories important for this book. The ordering of this subchapter might seem counterintuitive, but the goal is to establish a clearer understanding of the TB.

The Difference Between Time Banking and a Time Currency

TB and time currency are not synonyms. The concept of a currency of time is often used in various other contexts (Schroeder et al., 2011), for example in LETS systems such as Ithaca Hours (IthacaHours, 2013), but TB is specific system which has particular properties, as explained in the following subchapters. The focus of a TB might seem purely economic, but the transactional aspect is merely a vessel which brings other (usually positive) impacts to society and the development of trade and exchange in the place of its existence. TBs are also considered as catalyzers that can help to rebuild communities socially by creating "networks of reciprocity" (Gregory, 2009). Moreover, existing research also points out that TB can generate not only economic, but also social and political citizenship by facilitating reciprocal relationships (Seyfang, 2006). The TB differs in the fact that the economic concept of time currency is secondary. It only facilitates the co-production. This difference will be more and more clearer through the book. Where, for example, time currency based LETS aim economic development in an area, TB aims to facilitate societal problem solving and social innovations by well-known and transparent concept of trade/exchange. In this book we will not consider other time currency based economic systems than TB.

What Is a Time Bank?

The TB is a decades-old approach to utilizing the dormant aspects of market and human activities which cannot be sold on the labor market. During that time, it has spread from United States of America, its country of origin, to other countries around the world. Although in some sources (Miller, 2008) information can be found that the world's first TB was started in Japan and was founded in seventies by Teruko Mizushima, the idea started to spread globally approximately ten years later when E. S. Cahn invented the Time Dollar in US in the eighties (Cahn and Rowe, 1998).

TB is a socio-economic concept in its essence based on use of a time unit as a currency. However, unlike a monetary system it is based on equality and reciprocity (Clement et al., 2017). This means that instead of financial profit other benefits are perused in the sense of the utilization of unused resources, networking which cultivates social capital, and innovations. TB is a third-sector initiative that establishes principles of co-production and offers a model that enables users to be actively involved with other stakeholders in the delivery of services (Glynos and Speed, 2013). Understanding a TB in a particular environment, i.e. its adaptation to the current needs of a community, determines whether the social or the trade aspect is prevalent in a particular TB system.

The TB concept faces many challenges, and when it comes to various cultural environments, it is often adapted to reflect a particular regional reality (Valek, 2013a). Regional focus is one of features which define any TB, as it is focused on tackling locally and regionally based issues. Nowadays there are many TBs around the world, with their number constantly changing, thus it is quite difficult to determine the precise amount of them (Blanc, 2011, Schroeder et al., 2011). Due to these adaptations and the instrumentalization of TB for various purposes, TBs have been deemed to be "chameleonic" (Collom et al., 2012) and hard to identify. This is related not only to the precise number of TBs in the world, but consequently the number of people involved in them (Valor and Papaoikonomou, 2016). Nevertheless, whatever the adaptation, the focus is on locating ways to use skills, knowledge, services, etc. which people and organizations possess but are not using, as well as to provide a tool (Clement et al., 2017) by which both people and organizations can help themselves rather than expect support from authorities.

The TB is a complementary economic system, although sometimes it might be described as alternative (Clement et al., 2017, Valor and Papaoikonomou, 2016). This difference has already been explained above. In this book we consider TB to be complement existing in parallel to a mainstream monetary economy. And because TBs use time as a kind of currency, they can be deeply identified as a complementary currency. Nevertheless, members usually participate in a TB not for financial reasons e.g. to save money, but to gain other advantages from the system, such as learning, networking, social capital, etc. (Valor and Papaoikonomou, 2016).

Also, returning to the idea of a currency as an information system (Lietaer, 2001), the tracking of exchanges, or TB bookkeeping, among a constituency not only demonstrates a pure record of exchanges, it can actually show in detail the contributions of people and organizations toward building their community, growing social capital, tackling possible societal and business problems, and much more. This feature gives a TB system added value, as all exchanges can be very simply monitored and evaluated (Valek, 2018).

Core Values of Time Banking

TB was founded on the original four core values of Edgar S. Cahn, to which a fifth has been added more recently. These core values define a TB: (Granger, 2013, Boyle and Bird, 2014, Cahn, 2000, Ozanne, 2010, Valek, 2018):

1. **We Are all Assets:** Time Banking values people and recognizes that everyone has something special to offer others - knowledge, skills, resources, time, etc. Every human being has the capacity to be a builder and contributor as an individual or as a part of an organization (Granger, 2013). Thus, a service offered in a TB could be almost anything, a condition which opens up opportunities for exchange – making offers and requests – which are not "marketable," but have value for other members of the TB (both people and organizations).
2. **Redefining Work:** Redefining work means that members are rewarded for any kind of work, whether it would be considered financially profitable or not by market economy standards, as touched upon above. Work in this sense does not necessarily mean a "job."
3. **Reciprocity:** The exchange works in a two-way process. Members can both offer and request at the same time and both earning TC and spending TC brings positive feedback.
4. **Social Capital:** The creation of social capital is very important for any community, whether it consists of individuals or a knowledge-based organization (Nonaka, 1991, Valor and Papaoikonomou, 2016). Social capital, among other things, solves problems for individuals regarding both unemployment (Carnero et al., 2015) as well as learning. The formation of social capital facilitates innovation and further development of a society (Ozanne and Ozanne, 2016).
5. **Respect:** Every human being matters: as a mutual understanding of what people and organizations do, this is a key element in the development of positive relationships among entities in a region. A higher level of trust leads to the possibility of further innovation by the facilitation of common projects (Lehaney, 2004).

Typical uses of TB would include projects in community cohesion and neighborhood development (Cahn, 2000, Cahn, 2001, Cahn and Rowe, 1998, Marks, 2012, Collom et al., 2012), building social capital (Naughton-Doe, 2015, Molnar, 2011), the rehabilitation and revival of a local economy (Kennedy et al., 2012, Clement et al., 2017, Hallsmith et al., 2011, Martignoni, 2012), restorative justice (Gregory, 2012, Marks, 2012), healthcare (Boyle and Bird, 2014, Molnar, 2011, Lasker et al., 2010) with typical target groups ranging from ordinary people to unemployed, youth, elderly, prison inmates and businesses to institutions of local and regional government (Collom et al., 2012, Papaoikonomou and Valor, 2017, Whitham and Clarke, 2016, Collom, 2008, Clement et al., 2017, Valor and Papaoikonomou, 2016, Boyle, 2014, Gregory, 2012, Marks, 2012, Molnar, 2011, Cahn, 2001, Cahn and Rowe, 1998, Boyle and Bird, 2014, Ryan-Collins et al., 2008, Naughton-Doe, 2015, Timebanking UK, 2005, Timebanks USA, 2017, Carnero et al., 2015).

Time Banks and Business

Much like the business exchange, a CC is supported by a monetary based system; TB also involves businesses in its operation (Marks, 2012). These include models of TB termed person-to-agency and agency-to-agency, terms which will be explained in depth in the section on TB Taxonomy introduced later in this book. At this point, it can be stated that TBs offer possibilities to businesses to utilize free capacities of space, machinery, workforce, skills, knowledge as well as in many other areas which would otherwise be unused, thus in turn gaining counter-value which helps to raise the overall performance of a business. This benefit arises largely because of the involvement of human values in a work process rather than relying on typical "subject to object" communication (Smutek, 2017) in which employee is viewed only as a performing gear in a business machine rather than a multiple-layered systemic element. There are various ways how a business can take part in a TB. Aim of a business is mostly to gain a profit, therefore its motivation to participate in a TB would be often different from ordinary people. Although, in cases of social enterprises (Social Enterprise UK, 2017, Ceske socialni podnikani, 2018) the main focus is on the mission of the organization rather than profit. Main motivation for involvement of businesses into a TB is utilization of unused assets, and possibly skills and knowledge of employees. Such situation happens if a business needs something which it cannot afford, and/or has various resources which are not marketable for money. The second possibility is to offer own services or products at the time and space where customers lack money, but there is an explicit demand. This might happen in areas struck by an economic crisis, but also in economically developed ones. The basic catalyst allowing harmonic operation of both businesses and people in a TB environment is an offer and demand which cannot be met by money. To clarify we can introduce an example:

A person offers a service of shopping for elderly. Although there is high demand, there are no funds to pay for it in money. Thus, the person earns Time Credits. A local café offers refreshments, but at certain hours it lacks customers. During that time, it offers refreshments for Time Credits. Our person who did shopping for elderly can go there. Owner of café has customers instead of empty café, and earned Time Credits can be spent on a part-time cleaner on pension or on legal advice services related to taxation from a company across street which deals in accountancy. This accountancy company has at the moment lack of work, so it lets one of their accountants to provide advisory for Time Credits, as he/she is not as busy as he/she usually is (which is much better idea than dealing with hiring/firing and staff fluctuation in general). The accountancy company needs to provide trainings for their accountants and to rent a training room periodically. They use their Time Credits to pay for a training room provided by a local youth club, which has state-of-the-art computer training room for their coding classes. In this way, the network progresses to connect all possible actors in the area. All services and assets offered are unused and it is impossible or very complicated to offer them for money. By not using them a part of local economy and people capacities are left out of the economy. Without a TB facilitating these connections and exchanges, they would stay unused. This fact is also basis for arguments against taxation of the TB as it opens a part of exchange which would never happen through market economy. Nevertheless, there are more reasons underlined in next paragraph.

Controversies Regarding Law and Taxes

At the general level, sharing economy has problems with compatibility existing legal systems. Majority of our laws created over the last decades to manage economic relationships were becoming increasingly polarized or exploitative (Kassan and Orsi, 2012). A common critique of the TB system comes from the perspective of taxation. Sometimes it is argued that TBs are a tax avoidance and do not fit into a legal framework. The counter to this opinion is clear from an explanation of the very basis of the TB system; a TB is not created to simply avoid taxes, but it is actually an altogether different entity, as it is not related to money at all. The first important feature is the fact that many exchanges which take place in a TB would never have happen without it. A second potential argument against the TB is one of valuation, as it is almost impossible to tax an exchange which is not valued in money. In any case, taxation would place a financial burden on a non-financial system, thus destroying it in the process (Valek, 2013b). In addition, a TB aims to solve a societal problem and is one of least expensive tools for doing so. In other words, the success of a TB lies in an understanding by a government that TB can open up completely new possibilities, thus a small investment can actually produce great savings of money and effort toward co-productive problem solving. A government loses nothing, as exchanges facilitated by a TB would never have happened without it, so there is actually nothing to tax. There are countries where TB is not understood at all as well as ones where it is even supported by government (Valek, 2015, Gregory, 2012). For example, in the USA already by the eighties the Internal Revenue Service had exempted TB from taxation (Cahn, 2000), as later occurred in the UK (Gregory, 2012). In the UK TB is even considered by the Department for Work and Pensions to be an active approach in response to unemployment, i.e. an alternative by which a unemployed person can participate in a TB as a part of his/her job seeking efforts; but must still be search for a paid job in the process (DWP, 2015).

THEORETIC BASE TO TIME BANKING AND THE SOCIO-ECONOMIC CONTEXT

Link to Economic Theory

The first links to theory which relate to the values and function of TB in society can be found in certain works by early economists. The very first origins come from classical Political Economy, with traces to be found even in certain texts by David Ricardo in which he considers possibilities regarding a currency which would have a unchangeable value over time (Ricardo, 1934). Ricardo argues at the time that no such currency exists. This, of course, is true unless a currency is based on time itself, as time does not change value over time, thus a time-based currency whose value changed over time would be a tautology. This extrapolation from Ricardo's position from 200 years ago demonstrates the proof of the value stability of a time currency.

John Stuart Mill's Political Economy as well as the theory of Utilitarianism of Jeremy Bentham (Mill, 2011) can shed more light and bring clarification. Utilitarianism describes the need of an individual to maximize pleasure and minimize pain as much as possible (Mill, 2011). This seemingly simple dichotomy is itself based largely on equality (Mill et al., 1962), i.e. as all people are equal in achieving these needs and this is the means by which utility is defined, or, to be precise, the gain of utility. This point is further developed by J. S. Mill when he differentiates utility from base urges to satisfy primitive pleasures (Mill et al., 1962), thus placing utility into a context of the procurement of use on an equal level throughout society. In other words, individuality is important (Mill, 1881) and is in fact the driving force of society, including the economy, but this drive is based on equal and positive terms (Valek, 2018, Valek, 2016b). In public policy, the focus is placed on the putative "greater good," i.e. creating a condition which improves the lot of a certain group of people and worsens the situation of another, all the while maintaining the balance in favor of the good (improvement) (Sandu, 2012).

As for the trading concept related to TB, on page 265 of Mill's original Principles of Political Economy he argues that there is difference between price and value, with price being value demonstrated in money (Mill, 1900). In other words, prices do not always reflect value in general, especially value for an individual. Prices are set by various factors, such as costs, scarcity, complexity of production, etc., with values for individuals usually encompassing different measures. Another term outlined by Mill is "Value in Exchange" in (Mill, 1900). i.e. value which goods or services possess in an exchange between two parties. This value can be reset in terms of money, but only when there is another, more equal currency such as time. Time currency fits into this scheme much better than traditional money, as TC reflects the actual value for each person when both sides of a trade are set forth on completely equal and balanced terms (Valek, 2016b, Valek, 2018). In (Mill, 2011) more traces can be found which relate to TBs, e.g. a factor which can be called the "motivation to active participation," by which a utilitarian concept of society seems very close to Cahn's co-production (Cahn, 2000).

E. S. Cahn: Time Dollar and Co-Production

American civil rights lawyer E. S. Cahn presented the idea of TB in the nineteen eighties after a sudden and long-term illness caused him to feel marginalized in society as a burden or, as he wrote a "throw-away person" (Cahn, 2000). Cahn felt that he still he possessed many skills and knowledge that may be used by society and even in the labor market. During this period his idea of Time Dollars (Cahn and Rowe, 1998) was born, a concept which worked along the main core values mentioned above, thus creating the first TBs. The movement was successful and later spread to other countries. But the development of the concept did not stop there and it led to the implementation of an idea called co-production (Cahn, 2000, Cahn and Rowe, 1998, Clement et al., 2017), which, as we will learn later, is directly related to other theories, including knowledge management (Valek, 2018).

The market economy is profit-based, i.e. founded on products and services that can be traded. This would seem to leave most of the not-for-profit sphere of the economy behind, even though in reality this domain possesses the same resources as the sector motivated by financial profitability. Cahn argues that, for example, externalities should be a responsibility of the market economy which should also function outside of the seller-buyer paradigm (Cahn and Rowe, 1998, Cahn, 2001). As the main imperative of TB, co-production has been mentioned (Cahn, 2000, Cahn and Rowe, 1998, Clement et al., 2017), indicating the acceptance at all levels of society of the involvement of people as active problem-solving agents. The market economy in fact neglects many aspects of human life and society, including, notably, (many if not all) natural functions and capacities. The methods suggested here are not intended to replace the market economy, but simply to accept the non-market economy as an equal part of the economic paradigm, thus creating win-win economics (Cahn, 2000). Lietaer argues that, finally, we live in a world of constant flow and ever-present change, thus there is no reason to believe that economic concepts which we inherited from the industrial age will always serve us most optimally in subsequent periods and thus will remain unchanged forever (Lietaer, 2001).

Co-Production and Business Applications

The idea of co-production (Cahn, 2000, Cahn and Rowe, 1998, Clement et al., 2017) is often not implemented in business in a traditional profit-based understanding, at least not consciously. One exception might be social enterprises, which do not entail co-production in their definition, but follow similar ideas (Social Enterprise UK, 2017). As co-production is simply an approach by which people take responsibility for the development of their surroundings, environment and direction of societal development, this can be also accomplished in a business company. People organized within a company, whether managers or staff, can decide to take innovative steps in order to fully utilize the potentiality of skills, knowledge and assets, both on an individual level and on a company level. The triad of skills, knowledge and assets will be further designated as "SKA." The co-productive approach would require that people working within an organization take an active role in problem solving and vigorously seek solutions on their own when a negative situation occurs. This would take place of course within the framework of the organizational structure and rules of an organization. In relation to CCs, namely TB, this would also mean redefining the point of view of profit, i.e. that the organization is willing to share SKA not only for money, but also for certain non-monetary benefits for the company, a condition which ultimately brings benefits to everyone involved (Valek, 2018). This is strongly related to the theory of communities of practice, which will be explained in the methodology chapter.

Alan Fiske

The Relational Models Theory (RMT) of anthropologist Alan Fiske (Fiske, 1992) also can be related to TB background theory. RMT expresses the notion that human relationships use just four relational models in various combinations no matter the culture-specific environment (Fiske and Haslam, 2005). These models are: communal sharing (CS), authority ranking (AR), equality matching (EM), and market pricing (MP). These four models are structures out of which people construct, understand, evaluate, sanction and motivate most common activities (Fiske and Haslam, 2005).

The communal sharing model creates cohesion within a social group, i.e. the feeling of being part of a community where people feel they have something in common with differs between "us" and outsiders.

The authority ranking model assumes that there is a hierarchy, i.e. a structure which defines roles of a leader and followers. The leader has power, but also responsibility.

The equality matching model is about equality and reciprocity, even though a difference in the skills of people is acknowledged. People are equal, and exchanges of knowledge, time and possessions are fair and reciprocal.

The market pricing model of RMT defines the need of people to valuate work or goods in order to set differences, or better to say, proportions. The ratio could be monetary value, utility, efficiency, merit or anything else (Fiske and Haslam, 2005).

If we relate RMT to the TB sphere, we can find a clear connection. As RMT goes to the roots of society, it also touches the core values on which TB concept is based. At level of CS, TB focuses on building a community along with implementing co-production and a sense of common goal. AR is not emphasized in TB. There are no barriers, however, in the creation of a hierarchy, for example, according to skill level, contribution to the community or regarding the amount of exchanges. At minimum, there is usually coordinator which connects people, creating a very horizontal hierarchical model. EM is very typical for TB and it is actually one of basic aspects. Equal exchange based on reciprocity, which is based in turn on the various skills and knowledge of the people involved can be related directly to TB core values. MP maps also to the TB concept, but not in monetary value. The exchanges are valued by the utility which they bring to people. The MP utility value approach here is very similar to utilitarian theory, mentioned above. In essence this actually means that RMT is a very core theory of society, because if we compare these two theories, Utilitarianism and RMT have a very similar common grounding and, if we go deep enough to basics of a structure of a society, at bottom we will find the structure set by RMT (Valek, 2018).

Time Bank and Information Technologies

The situation in a TB environment related to information technologies, as is the case with the rest of the research background, has remained until now unexplored. At the moment there should be around about thirty IT solutions for TB have been designated (Boyle and Bird, 2014). This "should be" means that very often a group of people or an organization which decides to found a TB has conducted no preliminary research and develops its own software (Valek, 2016a). Three important models are Time & Talents from the organization hOurworld, Community Weaver from Timebanks USA and Time Online provided by Timebanking UK. These software initiatives are mostly based on a variant of content management solutions, such as

WordPress or Drupal. Another possibility for Time Bankers is the software Cyclos (Cyclos Software, 2016), which can be used as a platform for various kinds of Complementary Currencies, including TB. Cyclos must be purchased in its 4.0 version, but it can be used free of charge by a social organization if social benefits and non-profit aims have been demonstrated; Cyclos 3.0 is offered as open source (although support was ended in 2013), so it can help enthusiasts to create their own Complementary Currency trade system (Valek, 2016a). More or less stable, or more or less functional, an IT solution is now a part of any established TB. Software is also used when a TB decides to go the way of monetarized TC.

A task of a TB software is to provide a virtual space for exchange, usually in a form of an offer and request "notice" board. It often allows to evaluate the exchange and provide feedback about the service. Usually something as "news" page where coordinators often promote various social activities and some requests and offer is also included. Common practice is that to "log" a service/TC a consent of both sides is necessary, so it guarantees that service was provided in the designated duration and both sides agree.

Unfortunately, a deeper analysis of more than few TB software named above is a task for separate research. When we talked above about "chameleonic" feature of TBs, this is even more true for an IT system they use for operation. And as groups of people are funding TBs, and sometimes developing own solutions valid in their local context, without any registration or record keeping, it is almost impossible to access information about them including their IT solution.

REFERENCES

Benes, J. & Kumhof, M. (2012). *The Chicago Plan Revisited.* Academic Press.

bitcoin.cz. (2016). *Co je to Bitcoin?* [What is Bitcoin?]. Available: http://blog.bitcoin.cz/co-je-to-bitcoin/

Blanc, J. (2011). Classifying "CCs": Community, complementary and local currencies' types and generations. *International Journal of Community Currency Research, 15*(D), 4–10.

Boyle, D. (2014). *The Potential of Time Banks to support Social Inclusion and Employability. JCR Scientific and Policy Reports, Report EUR 26346 EN.* European Commission.

Boyle, D., & Bird, S. (2014). *Give and Take: How timebanking is transforming healthcare*. Timebanking UK.

Cahn, E. S. (2000). *No more throw-away people: the co-production imperative*. Washington, DC: Essential.

Cahn, E. S. (2001). On LETS and Time Dollars. *International Journal of Community Currency Research*, *5*, 1–4.

Cahn, E. S., & Rowe, J. (1998). *Time dollars: the new currency that enables Americans to turn their hidden resource--time--into personal security & community renewal*. Chicago: Family Resource Coalition of America.

Carnero, M. A., Martinez, B., & Sanchez-Mangas, R. (2015). Explaining transactions in time banks in economic crisis. *Applied Economics Letters*, *22*(9), 739–744. doi:10.1080/13504851.2014.975323

Caroll, J. M., & Belloti, V. (2015). Creating Value Together: The Emerging Design Space of Peer-to-Peer Currency and Exchange. In Collaborating through Social Media - CSCW 2015. Vancouver, BC, Canada: ACM.

CCRC. (2016). *Complementary Currency Resource Center*. Available: http://complementarycurrency.org/

Ceske Socialni Podnikani. (2018). *Czech social enterpreneurship - principles and definition*. Available: http://www.ceske-socialni-podnikani.cz/socialni-podnikani/principy-a-definice

Clement, N., Holbrook, A., Forster, D., Macneil, J., Smith, M., Lyons, K., & Donald, E. M. (2017). Timebanking, Co-production and Normative Principles: Putting Normative Principles into Practice. *International Journal of Community Currency Research*, *21*, 37–52.

Collom, E. (2008). Engagement of the Elderly in Time Banking: The Potential for Social Capital Generation in an Aging Society. *Journal of Aging & Social Policy*, *20*(4), 414–436. doi:10.1080/08959420802186282 PMID:19042555

Collom, E., Lasker, J. N. & Kyriacou, C. (2012). *Equal Time, Equal Value: Community Currencies and Time Banking in the US*. Academic Press.

Cyclos Software. (2016). Available: http://www.cyclos.org/products/

DWP. (2015). Department of Work and Pensions Statement on Timebanking - November 2015. DFW Pensions.

EEA. (2018a). *Members*. Available: https://entethalliance.org/members/

EEA. (2018b). *Working groups*. Available: https://entethalliance.org/working-groups/

Fiske, A. P. (1992). The 4 elementary forms of sociality - Framework for a unified theory of social-relations. *Psychological Review*, *99*(4), 689–723. doi:10.1037/0033-295X.99.4.689 PMID:1454904

Glynos, J., & Speed, E. (2013). Varieties of co-production in public services: Time banks in a UK health policy context. *Critical Policy Studies*, *6*(4), 402–433. doi:10.1080/19460171.2012.730760

Granger, P. (2013). *Valuing people and pooling resources to alleviate poverty through Time banking. Lewisham*. London: Rushey Green Time Bank.

Gregory, L. (2009). Spending Time Locally: The Benefits of Time Banks for Local Economies. *Local Economy*, *24*(4), 323–333. doi:10.1080/02690940903026852

Gregory, L. (2012). Time and punishment: A comparison of UK and US time bank use in criminal justice systems. *Journal of Comparative Social Welfare*, *28*(3), 195–208. doi:10.1080/17486831.2012.742452

Hallsmith, G., Lietaer, B. A., Lovins, L. H., & Meadows, D. L. (2011). *Creating wealth: growing local economies with local currencies, Gabriola, B.C.* New Society Publishers.

hearth.net. (n.d.). *The gift economy*. Available: www.hearth.net

Holmgren, D. (2011). *Permaculture: Principles & Pathways Beyond Sustainability*. East Meon, Permanent Publications Hyden House Ltd.

Ithacahours. (2013). *Main Page*. Available: http://www.ithacahours.com/

Jevons, W. S. (1875). *Money and the mechanism of exchange*. London: H.S. King & Co.

Kassan, J., & Orsi, J. (2012). The LEGAL Landscape of the Sharing Economy. *Journal of Environmental Law & Litigation*, *27*, 1.

Kennedy, M., Lietaer, B., & Rogers, J. (2012). *People money: The promise of regional currencies*. Axminster, UK: Triarchy Press.

Lasker, J., Collom, E., Bealer, T., Niclaus, E., Keefe, J. Y., Kratzer, Z., ... Perlow, K. (2010). Time Banking and Health: The Role of a Community Currency Organization in Enhancing Well-Being. *Health Promotion Practice*. PMID:20685912

Lehaney, B. (2004). Beyond Knowledge Management. Idea Group Publishing.

Lietaer, Arnsperger, Goerner, & Brunnhuber. (2012). *Money and sustainability: the missing link*. Club of Rome-EU Chapter to Finance Watch and the World Business Academy.

Lietaer, B. (2001). *The Future of Money*. London: Century, Random House Group Limited.

Lietaer, B. A., & Dunne, J. (2013). *Rethinking money: how new currencies turn scarcity into prosperity*. San Francisco, CA: Berrett-Koehler.

Marks, M. (2012). Time banking service exchange systems: A review of the research and policy and practice implications in support of youth in transition. *Children and Youth Services Review*, *34*(7), 1230–1236. doi:10.1016/j.childyouth.2012.02.017

Martignoni, J. (2012). A new approach to a typology of complementary currencies. *International Journal of Community Currency Research*, *16*, A1–A17.

Martignoni, J. (2015). *Monetary velocity in a systemic perspective From Macroeconomics to Currency-Thinking. In 1st Annual Conference of Monetary Research Center Monetary Policies In The Balkans*. Sofia, Bulgaria: The Future of the Euro and Eurozone in the Balkan Region.

Mill, J. S. (1881). *O individualite, Praha, V kommissi knihkupectvi Jarosl.* Pospisila.

Mill, J. S. (1900). *Principles of Political Economy*. London: Longmans, Green, and Co.

Mill, J. S. (2011). *Utilitariarismus*. Praha: Nakladatelstvi Vysehrad, spol. s r.o.

Mill, J. S., Bentham, J., & Austin, J. (1962). *Utilitarianism*. Glasgow, UK: William Collins Sons & Co. Ltd.

Miller, J. (2008). Teruko Mizushima: Pioneer Trader in Time as a Currency. *Intersections, 7*.

Molnar, S. (2011). Time is of the essence: The challenges and achievements of a Swedish Time Banking initiative. *International Journal of Community Currency Research, 15*, 13-22.

NakamotoS. (2009). Available: https://bitcoin.org/bitcoin.pdf

Naughton-Doe, R. (2015). *An evaluation of timebanking in England: What can timebanks contribute to the co-production of preventive social care?* (Dissertation). University of Bristol.

Nonaka, I. (1991). *The Knowledge-Creating Company*. Harward Business Review.

Ozanne, L. K. (2010). Learning to exchange time: Benefits and obstacles to time banking. *International Journal of Community Currency Research, 14*, A1–A16.

Ozanne, L. K., & Ozanne, J. L. (2016). How alternative consumer markets can build community resiliency. *European Journal of Marketing, 50*(3/4), 330–357. doi:10.1108/EJM-12-2014-0802

Papaoikonomou, E., & Valor, C. (2017). The Institutionalization of Alternative Economies: The Processes of Objectification in Time Banks. *Journal of Macromarketing, 37*(2), 167–179. doi:10.1177/0276146716672286

Ricardo, D. (1934). *Zásady politické ekonomie a zdaňování [Principles of Political Economy and Taxation]*. Praha: Prometheus.

Rifkin, J. (1995). *The End of Work: The Decline of the Global Labor Force and The Dawn of the Post-Market Era*. New York: G. P. Putnam's Sons.

Ryan-Collins, J., Stephens, L. & Coote, A. (2008). *The new wealth of time: How timebanking helps people build better public services*. London: New Economics Foundation.

Sandu, A. (2012). *Appreciative Ethics*. Iasi, Romania: Lumen publishing house.

Schroeder, R., Miyazaki, Y. & Fare, M. (2011). Community Currency Research: An analysis of the literature. *International Journal of Community Currency Research, 15*(A), 31-41.

Seyfang, G. (1999). The Euro, The Pound, And The Shell In Our Pockets: Rationales for Complementary Currencies In A Global Economy. *New Political Economy, 5*(2), 227–246. doi:10.1080/713687774

Seyfang, G. (2006). Harnessing the potential of the social economy? Time banks and UK public policy. *The International Journal of Sociology and Social Policy*, *26*(9/10), 430–443. doi:10.1108/01443330610690569

Seyfang, G., & Longhurst, N. (2012). *Money, money, money? A scoping study of grassroots complementary currencies for sustainability*. Norwich: Science, Society and Sustainability Research Group.

Smutek, M. (2017). Life-world and System in Social Services. Hradec Days of Social Work: Social Work as a Means of Promoting Social Change. Hradec Kralove, Czech Republic.

Social Enterprise UK. (2017). *Social Enterprise.* Available: http://www.socialenterprise.org.uk/advice-services/sector/social-enterprises

The Economist. (2013, November 30). Digital money: The bitcoin bubble. *The Economist.*

Timebanking UK. (2005). A bridge to tomorrow: time banking for baby boomers. Gloucester, UK: TimebanksUK.

Timebanks USA. (2017). *About Timebanks USA*. Available: https://timebanks.org/timebanksusa/

Transferwise. (2016). Available: https://transferwise.com/

Valek, L. (2013b). Time banks as systematic approach in solution of economical problems on local scale: The Czech Republic situation. In *Systematic Approaches 2013: Systems thinking and global problems of the world*. Prague: University of Economics, Publishing Oeconomica.

Valek, L. (2013a). Risk Factors and Issues Which Might Arise While Implementing and Running a Time Bank: Learning from Success and Failure. In *22nd International-Business-Information-Management-Association Conference on Creating Global Competitive Economies: 2020 Vision Planning and Implementation.* Rome, Italy: Int Business Information Management Assoc-Ibima.

Valek, L. (2015). The difference in understanding of time banking in various contexts. In *6th LUMEN International Conference: Rethinking Social Action. Core Values 2015*. Iasi, Romania: LUMEN.

Valek, L. (2016a). Open Ways for Time Banking Research: Project Management and Beyond. *International Journal of Human Capital and Information Technology Professionals*, *7*(1), 35–47. doi:10.4018/IJHCITP.2016010103

Valek, L. (2016b). Time Banking and its place in economic education. *Journal of Economics and Economic Education Research*, *17*, 9–15.

Valek, L. (2018). Three New Directions for Time Banking Research: Information Management, Knowledge Management, and the Open Source Model. In V. Ahuja & S. Rathore (Eds.), *Multidisciplinary Perspectives on Human Capital and Information Technology Professionals*. Hershey, PA: IGI Global. doi:10.4018/978-1-5225-5297-0.ch017

Valor, C., & Papaoikonomou, E. (2016). Time Banking in Spain: Exploring their Structure, Management and Users' Profile. *Revista Internacional de Sociologia*, *74*(1), 14. doi:10.3989/ris.2016.74.1.028

White, L. H. (2015). The merits and feasibility of returning to a commodity standard. *Journal of Financial Stability*, *17*, 59–64. doi:10.1016/j.jfs.2015.03.003

Whitham, M. M., & Clarke, H. (2016). Getting is Giving: Time Banking as Formalized Generalized Exchange. *Sociology Compass*, *10*(1), 87–97. doi:10.1111oc4.12343

Chapter 2
Methodological Approach to Development of the Universal TB Model

ABSTRACT

While the previous chapter can be considered as purely theoretical, this chapter focuses on the methodology. It introduces the details associated with methods, techniques, and approaches whose application led to the main output of this book. Firstly, an extensive literature review that provides background for discussion of available resources and existing knowledge background is provided. Semi-structured interviews help to acquire insights of insiders. Both methods help to identify the main issues and problems, non-existence of unifying systemic definition included. This chapter also provides details to applied statistical analysis and systems approach which represents a framework in which the whole TB analysis and synthesis are conducted. Special attention is paid to the design and construction of various types of systems and description of TB as a soft system.

APPLIED METHODS

Due to the complexity and multidisciplinarity of the research, various methods were used, starting with a literature review researching available resources, progressing to a picture of the background in research, going on to semi-structured interviews (with stakeholders around TBs and with members),

DOI: 10.4018/978-1-5225-6974-9.ch002

questionnaires (to learn more about particular TBs), internet searches, a scientific database research, statistical analysis and finally an in-depth analysis. Further methods include comparison, the Spiral of Knowledge also known as SECI model (Nonaka, 1991), Systems Analysis (Bureš, 2011), Soft System Methodology (Checkland, 1993) as well as a synthesis of all these. A strong correlation among Information and Knowledge Management and Information Technologies was also found, with relevant theories researched and implemented.

Prior to an analysis of TB as a system, or more precisely a soft system, the precise nature of TB should be explored. This includes defining what it is, identifying it as a system and soft system, and determining a methodology for analysis. This line of exploration was followed below.

Literature Review

The literature review had to entail a very extensive reach into many disciplines and sciences. This includes, but is not limited to, economics (Holmgren, 2011, Clement et al., 2017, Lietaer et al., 2012, Jevons, 1875, Samuelson and Nordhaus, 2005, White, 2015, Lietaer and Dunne, 2013, Mill, 1900, Boulding, 2004), knowledge management (Hildreth et al., 2000, Lehaney, 2004, Lesser and Storck, 2001, Nonaka, 1991, Otcenaskova et al., 2012, Shreiber et al., 2002), information management (Anthes, 2016, Bureš and Tučník, 2014, Linderman et al., 2005, Maes, 1999, Maes et al., 2000), complementary economic systems (Blanc, 2011, CCRC, 2016, Clement et al., 2017, DeMeulenaere, 2008, Hale et al., 2004, Hallsmith et al., 2011, Kennedy et al., 2012, Martignoni, 2012, Seyfang, 1999, Seyfang and Longhurst, 2012a, Seyfang and Longhurst, 2012b, Schroeder et al., 2011, Economist, 2013, Lietaer et al., 2012, Lietaer, 2001), general systems theory and systemic approaches (Boulding, 2004, Bureš, 2011, Checkland, 1993, Checkland, 1999, Kim, 2000, Kim and Lannon, 1997, Rapoport, 1967, Senge, 2006, Skyttner, 2001, Skyttner, 2005, Smutek, 2017, Wiener, 1951), sociology and anthropology (Beamer, 1992, Fiske, 1992, Fiske and Haslam, 2005, Sandu, 2012), psychology (Jordan, 1968, Ilin, 2013). Some results of these findings have already been outlined above. TB is very broad topic and in each of these fields it would be possible to conduct extensive research. For this book, it was necessary to explore briefly all possible connections to various fields, but the main direction explored is systems engineering. To maintain focus and coherence throughout the book, only the most relevant fields were noted, and only the most pertinent resources are cited where appropriate.

Because the literature on the subject was scarce to say the least, a new approach had to be devised. As indicated in the introduction, the extant literature on topic generally analyzes functional TBs from various points of view regarding usability and impact on society, but no systemic approach was found, nor any relation to information and knowledge management except a brief mention of TB as a knowledge base (Valor and Papaoikonomou, 2016). The main resource for CES is International Journal of Community Currency Research (IJCCR), the central point for all the research in the field (IJCCR, 2017).

The literature research was begun by a web search which traced a few tracks regarding TB in various countries, but mostly in the United States of America (USA) and United Kingdom (UK). In both the USA and UK there is long tradition of TB, although each slightly differs in its presentation (Gregory, 2012). Both approaches have been social and aimed at helping society in terms of a local economy, but the USA approach is focused on neighbor help (TBUSA, 2016, Cahn, 2000), whereas the UK approach is more straightforwardly "social." In the UK, TB is a social service (Timebanking UK, 2015, Timebanking UK, 2005, Naughton-Doe, 2015, Boyle and Bird, 2014, Boyle, 2014), and in the US, TB is an alternate way of solving societal problems using a specific tool (hOurworld.org, 2016a, Clement et al., 2017, Collom et al., 2012). These are just two examples of adaptations, but they emerge as very important, as they have been adopted by most developed countries in the world relative to the length of use of the TB concept. Nevertheless, each TB is slightly different depending on how it is instrumentalised (Boyle, 2014, Valek, 2015, Valor and Papaoikonomou, 2016). This matter is elaborated on below in the section focused on definition of the TB.

The natural next step of the research was use of scientific databases such as Thomson Reuters Web of Science, Scopus, EBSCO, Science Direct, and to a lesser extent Google Scholar. The following keywords were utilized: Time Bank, Time Banking (in various forms), cooperative economy, sharing economy, complementary economy as well as additional adaptations related to forms of non-mainstream economies. This query returned only a limited amount of usable scientific research in the field. Those which contained scientific value are widely cited in this book along with more practically aimed resources from real-life working TBs. In total there are 31 viable resources.

This scarcity of suitable resources was confirmed by other researchers, such as Naughton-Doe (Naughton-Doe, 2015), Clement, et. al. (Clement et al., 2017) as well as Valor and Papaoikonomou (Papaoikonomou and Valor, 2016, Papaoikonomou and Valor, 2017, Valor and Papaoikonomou, 2016), as introduced in literature review section.

Furthermore, the continuing research demonstrated that there are countless other study adaptations (Valek, 2016a, Valor and Papaoikonomou, 2016, Clement et al., 2017, Naughton-Doe, 2015) shifting towards the project management area (Valek, 2014). At this point, the situation in the Czech Republic had to be assessed; here the research showed that there have been several tries at various levels to set up a TB in this country, but they usually failed for several reasons, the main cause being mingling of the TB's economic base with a monetary alternative, rather than accepting the system as a complement which supports the economic reality at its weak points (Valek, 2013b, Valor and Papaoikonomou, 2016). In other words, the reason was that the concept was applied as an alternative to the monetary system installed from top to bottom without previous research; i.e. with a focus on a new utopist approach. TBs in the Czech Republic have mostly been founded as an alternative kind of economy (Valek, 2013a, Clement et al., 2017, Valor and Papaoikonomou, 2016) rather than as a systemic solution to a particular problem with extensive and focused system-wide thinking. In order to successfully apply the concept in any sphere, whether in a non-profit or a business, it is necessary to conduct research and to select an appropriate approach for the current conditions. On the other hand, a universal approach that can be applied everywhere is an approach that this book begins to describe.

Definition

A search for the ultimate definition of TB followed with the goal of establishing a solid basis from which to progress to the field of systems sciences. To our knowledge, TB has never been clearly defined; rather it has been perfunctorily described, much like in the section of this book "What is Time Bank?" More precisely, there is no unifying systems based definition. This is actually a common feature for most community currencies (Schroeder et al., 2011). Therefore, we never know exactly what TB system is, we can only learn what it does. But even these goals and results differ based on region, approach, variant of TB, etc.

The reason that no unifying definition is outlined in previous section is the lack of real research. One study found to be relevant uses various descriptions as a basis to explain particular TB or TBs in a relevant context. Thus, where we would expect to find a definition there are usually only descriptions. In addition, even if we consider these descriptions as defining, they lack a systemic background, often omitting or adding features without any recognizable pattern. This might seem satisfactory in the short run and in the context of small local researches, but an unequivocal definition would greatly contribute to the clarity of any TB related research. Of course, with a variety of points of view conflicting definitions emerge. In addition, the use and environment of a particular TB plays a role, but one key definition which would take into consideration all core elements and features of TB as a system should be developed.

The introductory sections of most of the relevant studies which contain a description of TB naturally almost always cite E. S. Cahn and his original idea of Time Dollars and five core values, often supported by an explanation of co-production phenomena (Cahn, 2000, Cahn, 2001, Cahn and Rowe, 1998, Clement et al., 2017). Only by these elements can we confirm that a study actually concerns TB. One example is Gill Seyfang's definition of TB based on core values: "Recognising people as assets and that everyone has skills to share; redefining work to include the unpaid 'core economy' of work in the neighborhood and community; nurturing reciprocity and exchange rather than dependency; growing social capital; encouraging learning and skills-sharing; involving people in decision making"(Seyfang, 2006). But this is still closer to a paraphrasing of core values than a real systemic definition, as it lacks a definition of the elements and processes involved in a TB system. More examples of defining TB in this descriptive way follow. The following are direct citations from articles, followed by citations from the websites of some TBs from various countries. Texts were translated when necessary.

Journal Articles

- Explaining transactions in time banks in economic crisis from Applied Economics Letters, 2015 Vol. 22, No. 9 by M. Angeles Carnero, Blanca Martinez and Rocío Sánchez-Mangas

Time banks were designed to create social networks and neighbourhood support by allowing participants to trade services among themselves without money. Each hour of community work is rewarded by the bank with one credit of a time-based local currency, independently of the service provided; therefore, everyone's time is worth the same. Participants use their time credits to purchase services from other users and/or store their earned credits to use them later. This implies a complete change in the valuation of labour that will not be based on market prices but on the individual value of time.

- **Time and Punishment:** A comparison of UK and US time bank use in criminal justice systems from Journal of Comparative Social Welfare Vol. 28, No. 3, October 2012, 195–208 by Lee Gregory

Time banks were designed in the 1980s, by the US civil rights lawyer Edgar Cahn, as a means to rebuild diminishing social networks and informal neighbourhood support that forms the core economy. As a community currency, time banks provide time dollars (time credits in the UK context) to individuals who donate time to help their communities, with each hour treated equally regardless of the activity involved. Each hour of volunteering is equal to one time dollar, which can ''purchase'' an hour of service/activity from their community or be used to access goods and services. For example, Toni uses one time credit to be driven to the shops and back by Alex, thus earning her one time credit. She then uses her time credit for piano lessons from Phil, who uses his earned time dollars to have Toni visit him through a befriending scheme. All of these interactions are recorded by the Time Broker, responsible for recording how people spend and use credits and recruiting new members to the time bank, thus increasing the range of skills and services available to members. Consequently, time dollars operate on a different value system to mainstream national currencies, which determines the value of activities based on scarcity and ability to create further money. Underpinning time banks is a theory of co-production. … This modification establishes four core concepts, forming the universal principles of time banking: treating people as assets capable of contributing positively and productively to a service; redefining work to accept a wider range of activities for their beneficial outcomes; reciprocity as the defining characteristic of the relationship between members; and time banks as a practical means of establishing social capital within a community, to ensure community cohesion and re-establish trust in local settings. Schemes that pursue a community purpose and principally aim to contribute to the "well-being, empowerment, autonomy and social exchanges of a given community

- **Time Banking Service Exchange Systems:** A review of the research and policy and practice implications in support of youth in transition from Children and Youth Services Review 34 (2012) by Michael B. Marks

Time banking begun by Edgar Cahn in the 1980s, is a unique transaction based system to create economic opportunities, mutual aid and assistance. It involves the creation of a social network of members who exchange services, with each member committing to both giving and receiving services. Time bank members can include youth, family members, local businesses, staff from participating organizations and organizations themselves. Members can provide services to one member of the time bank and receive services from another member in the network. For example, one member can cut another person's hair. The person who receives the haircut delivers groceries for an elderly person. Members can also exchange services with organizations that they are involved with as clients, such as youth providing mentoring or tutoring services to others or receiving such services from fellow clients. In time banking, each kind of service has equal value, whether it is running errands for a senior citizen or providing legal care for immigrants. Each hour of service equals a time dollar. Hours of service are entered into a computer bank for use when members need help or support from others.

- **Time Is of the Essence:** The challenges and achievements of a Swedish Time Banking initiative from International Journal of Community Currency Research 15 by Stefan Molnar

The human-rights lawyer Edgar Cahn invented time banking to promote four "core values" that he deemed are missing in today's society. These values are 1) the idea that all people are resources; 2) increased reciprocity; 3) increased social capital; and 4) a reevaluation of the definition of "work". Participants in the time banking system commit themselves to helping other participants or to developing their local communities. For every hour of work, the participant receives one time credit. The credits received by any one participant are placed in his or her own account in a kind of "bank," exemplified by a ledger book or an IT--system. In some cases, printed coupons are used. Time credits are withdrawn from the account when the participant receives a service or gets to participate in an activity arranged by someone else. All transactions are monitored and arranged by a person employed as a "time-broker". Although all time banking systems operate according to these basic principles, the transactions can be organized in many different ways. In the person-to-person model, individuals trade services with other individuals within the local community. In the person-to-agency model, which is based on the principle of co-production, the time-banking tool is set up within an organization such as a hospital, a school or an association as a means of motivating people to help reach the organization's goals. In the agency-to-agency model, time banking is used as a way to facilitate exchange and cooperation between organizations.

Time Banks and Time Bank Network Statements

- Tampa Bay Time Bank – a community skills Exchange (USA); 597 MEMBERS. 22,986 HOURS EXCHANGED at http://tampabaytime.org/

Tampa Bay Time Bank is a community skills exchange, also known as a "time bank." Time banking is a way to exchange services as a complement to the cash economy. In Tampa Bay Time Bank we trade skills gained from work, education, life experience, hobbies, and volunteering, while making friends and building community.

Time banking is egalitarian. All activities are valued, and all activities are valued equally. One hour equals one hour. Each hour of work is equal to every other, regardless of the activity. We earn credit by doing something for another member, then we can use the credit by having any other member do something for us.

- Time Bank Mahoning Watershed (USA) at http://tbmw.org/about-tbmw/

A Time Bank is a community of people who support each other. Simply spending an hour doing something for somebody in your community, you earn a Time Credit. Then you can use that Time Credit to buy an hour of a neighbor's time or engage in a group activity offered by a neighbor.

- Timebanking UK – Time Bank umbrella organization/network of about 300 Time Banks in United Kingdom (UK) at www.timebanking.org

Timebanking is a means of exchange used to organise people and organisations around a purpose, where time is the principal currency. For every hour participants 'deposit' in a timebank, perhaps by giving practical help and support to others, they are able to 'withdraw' equivalent support in time when they themselves are in need. In each case the participant decides what they can offer. Everyone's time is equal, so one hour of my time is equal to one hour of your time, irrespective of whatever we choose to exchange. Because timebanks are just systems of exchange, they can be used in an almost endless variety of settings.

- Argyll and Bute Time Bank network (UK – Scotland) at http://www.argylltsi.org.uk/?page_id=108

Time banking is a way for people and organisations to share resources, skills and knowledge and be rewarded for it – in time. For every hour of time you give, providing a service for another member, you receive one time credit (1 hour = 1 time credit). These time credits can then be spent within the time bank community on things that you may need help or support with. It works on the principle that everyone's time is equal, so one hour of my time is equal to one hour of your time, irrespective of whatever we choose to exchange. Because time banks are just systems of exchange, they can be used in an almost endless variety of settings.

- Rushey Green Time Bank – Lewisham, London (UK); 536 MEMBERS. 66,808 HOURS EXCHANGED at https://www.rgtb.org.uk/

Time Banking is a community development tool and works by facilitating the exchange of skills and experience within a community. Time Banking values everyone's time as equal, 1 hour equals 1 hour. For every hour you spend helping someone in your community you are entitled to an hour of help in return.

- Dobrobank (Ukraine) at http://www.dobrobank.com/home/presentation.html

All operations of the Bank of Time are held by participants on the basis of a voluntary exchange of "Goodhours." The bank is managed by the coordinator. Experts who check a certificate recording the performance of specific skills and abilities of clients (medical, construction, child care, etc.) are involved in the work of the bank. The number of bank members can range from several hundred to a few thousand. In order to be able to participate in the program, you need to be trained in communication, to demonstrate the quality of the skills and abilities and to present your Curriculum Vitae. Currently, a system is offered for the exchange of time using a database of skills and abilities. The coordinator shall be informed regarding the exchange of time between participants and the number of hours spent, following which the credit and debit of the respective participants will be entered into the record.

- Time Bank Nizhny Novgorod (Russia) at http://nnov.timebank.ru/about

The Bank of Time is a socio-economic model based on the principles of mutual assistance and the initial equality of all participants and using non-market mechanisms. In the face of a lack of "classical" money, this step could help solve the problems of unemployment, unequal access to basic goods and services, social unsettledness and the revival of local communities.

- Bancos de Tiempo network (Spain) at http://www.bdtonline.org/

A Time Bank is a tool with which a group of people can create a social economic alternative. In a Time Bank, skills are exchanged between members without using money; only the hours of service provided and received are accounted.

- Zeitbank (Germany) at http://www.zeitbank.net/?page_id=7

People do something for other people (such as care or neighborhood help) and the hours are credited to the personal time account of ZeitBank network participants. For this, each participant can receive compensations in time hours or e.g. save his time credit as a pension. Participants who have no time hours on their account or offer no services themselves can buy hours (currently € 8.00 / hour).

ZeitBank is a virtual bank that manages and documents the time credits of the network participants using an internet booking system. In this sense, it works like an online banking system. For participants who are not online, ZeitBank organizes local branches, where each account administrator enters their bookings on the internet. These participants use booking checks which are signed by the service user and service provider.

Events Organization

To practically evaluate lessons learned from the literature, conduct semi-structured interviews and verify findings, two international events were organized based on the non-profit, non-governmental organization KURO Hradec Kralove z.s., of which the author is co-founder. The first event was conducted to evaluate possibilities regarding time banking on the European level. Among its other activities, this event offered the possibility to compare the status of TB in a number of European countries whose representatives were present and were able to share their diverse experiences. The event, which took place in the Russian city of Nizhny Novgorod in November 2014, consisted of a seminar organized in cooperation with two non-profit organizations, the previously mentioned KURO Hradec Kralove z. s. (Czech Republic), and SFERA (Russia). The project was funded by the Erasmus+ program of the European Commission. The seminar involved various individuals connected with various TB environments representing eight countries. Some of participating organizations were only in the early steps of time banking, which was nevertheless beneficial for this study defining the basic precepts of the TB concept. Six operational time banks were represented: Rushey Green Time Bank (UK), the network Timebanking.co.uk (UK), Bank Vremeni Nizhny Novgorod (Russia), Timebanking Romania (Romania), Charkiv people (Ukraine) and ComeUnaMarea Onlus (Italy). This event provided an excellent environment for comparisions of differences in particular systems and locations. Among the important attendees were Sarah Bird CEO of Timebanking UK; Philippe Granger from Rushey Green Time Bank, Lewisham, London; Tam Cassidy representing two Scottish TBs and Natalia Nikitina, founder of Nizhny Novgorod Time Bank (mentioned above connected with TBs in Russia). All other participants were coordinators or founders of their respective TBs. The next seminar was organized in Hradec Kralove, Czech Republic in January 2016. This time the topic was the development of a European network called Timebanking Europe (TBE) and a related focus on the inclusion of youth to TB. The base of participants was very similar, with only the student-founded and run Timebanking Romania having collapsed by then, but with the addition of the French organization TempOh and Finnish group Kakspy (a non-formal TB based in Helsinki). In addition, one guest was featured in the seminar as an expert, Terrence Daniels, one of developers of Time & Talents software and one of three lead members of hOurworld. These events had a strong impact on consolidating the TB research complied thus far by presenting a direct opportunity to verify theoretical findings against real-life experience within various countries.

Semi-Structured Interviews

In order to get a clear picture of how the TB concept is applied in reality, how it functions, what possible approaches exist, as well as what key elements are involved, it was necessary to obtain real information from people involved. Throughout the years, research meetings and semi-structured interviews (Svaricek and Sedova, 2014) have been conducted with people directly involved in TB, starting with members of TBs and coordinators, but also CEOs. The interviews were done either in person or by Skype with individuals in several countries, among which were the United States of America, where TB was founded, and the United Kingdom, where the TB movement is very strong, developed and still developing. Other nations represented include Russia, Ukraine, Italy, France, Finland and Turkey. This part of research was particularly useful, as it was determined that TB is always both adapted and adaptable to local conditions. Naturally, many interviews were conducted at the events described above, but many also during on-spot visits to the respective TBs. These findings provided a background for a general analysis of TB as a whole in its various adaptations to particular environments, underlining the universality of the concept if broken down into fundamental elements (Valek, 2015). This was only way to research specific differences and applications of TB since no literature is available. This time-consuming process of accessing experts' tacit knowledge was conducted over the whole period of the doctoral research.

The interviews included CEOs of Timebanking UK, Time Banks USA, the newly founded organization Timebanking Europe as well as organizations connected to these networks, along with their members and coordinators. At least from an applications point of view, the query can be said to contain most of the Time Banking world.

Knowledge Management Context

As will be introduced below, one of main functions of a TB is knowledge and skill sharing, with knowledge and skills used to perform a service exchanged. Knowledge sharing is one of most important features of a TB (Valek et al., 2014) and this happens in several ways. At first, TB provides people/members with a safe environment to find other people with similar interests who will gather around certain issues (along with people bound in standard organizations), much like communities of practice in a company (CoP) form around problems informally. A CoP is by one definition a "group of professionals informally bound to one another through exposure to a common class of problems, common pursuit of solutions, and thereby themselves embodying a store of knowledge" (Manville and Foote, 1996). This could be within a firm or any other organization in which a TB is included. The formation of CoPs has the same basis as the idea of co-production, introduced above as the main driving force of TB. Co-production is based on giving people the tools, or a "vehicle" (Clement et al., 2017) to be able to solve problems, so those with the appropriate skills, knowledge and interests, can naturally join the effort. These individuals will gather around the problem without much facilitation and start solving it themselves. Nevertheless, a framework would be handy in these conditions.

CoPs are in essence interest groups which gather informally around a problem to solve it, i.e. in the structured environment of an organization unsolved problems remain, thus people interested in solving those problems with the knowledge to do so gather naturally voluntary around an issue and resolve it. The only difference between interest groups of TB and CoPs is that CoPs form within an existing company, and interest groups (CoPs of a TB) form in an existing and working TB.

Of course, in TB both explicit and tacit knowledge is present. Explicit knowledge easily learned and transferred, for example knowledge contained in books. Tacit knowledge comes only by experiential learning and is gained by an individual over time (Shreiber et al., 2002). To "get out" the knowledge from the individual's mind is very complicated, but a process that must be undertaken so that others can learn from these skills and experience.

Learning from each other's knowledge within a TB is one of the motivating points for which to join. This raises the competences of all members, a possibility that many are attracted to, including organizations. This learning happens naturally just by the exchange of certain services, but also by other outputs and spin-off events and projects. This learning happens by people offering services, what they like and know how to do, and learning is a by-product of exchange of knowledge contained. For this reason, educational processes are fundamental part of the framework of a TB. This is in accordance with Nonaka's Spiral of Knowledge, where explicit knowledge can be extracted out of the tacit, as depicted below (Nonaka and Takeuchi, 1995).

The steps of the spiral are:

- **Socialization:** Seeing how others do it and learning from their knowledge, which happens when people meet and have the opportunity to experience the knowledge of another person, thus be shown how to do things. In TB there are services which are directly aimed at learning, but knowledge is transferred from one person to another also even in those which are not.
- **Externalization:** Now it is possible to formulate and describe the knowledge in a way which is understandable for others. Thus, part of the knowledge of the other person has been "extracted" and the learner is now ready to try new tasks and activities him/herself.
- **Combination:** To formulate how the newly gained knowledge may be used. In TB, this means developing new competences.

Figure 1. Spiral of Knowledge by Ikujiro Nonaka
Source (Nonaka and Takeuchi, 1995).

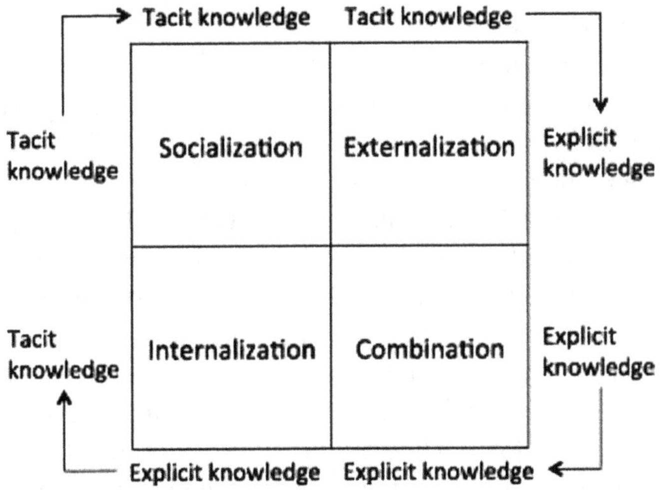

- **Internalization:** Using in application what has been learned. In TB, typically new services can be offered, new interest groups formed, new projects started, etc.

As there is not much formal information about TBs in general and most research has taken place by learning from others' tacit knowledge, the Spiral of Knowledge was also one of main research methods used to gain knowledge about world of TB, its contexts, work, applications, processes, etc. Explicit knowledge in form of literature is very limited, therefore the author had to either personally visit places where TBs operate and do most of the research on the spot or conduct research from a distance in contact with responsible people who possess the needed tacit knowledge. This entails the socialization aspect, which led to gaining contacts, itself a very time-consuming process. The later externalization entailed discussions and interviews, observation, study visits, etc., as described above. In this book this process is related to the introductory section, which unveils the context of TB. Combination is demonstrated in the following section, TB Taxonomy and Analysis. And finally, internalization is represented by the discussion and conclusion chapters, where new assumptions are formulated, and new directions of research proposed.

Statistical Analysis

Two datasets were obtained from two larger TBs in the UK during the research. Initially, the datasets are used to aid in research of Agent-based computational economics, (Tesfatsion, 2003, Bureš and Tučník, 2014) or ACE, which is used to simulate various economic realities. Subsequently this concept was enriched by the non-monetary component of Time Banking (Tucnik et al., 2016), where agent-based simulations have many potential uses.

As indicated above, the datasets are from two TBs in United Kingdom which are members of Timebanking UK (Timebanking UK, 2015), and they come from an older unit of software used for TB called Time Online. For the protection of personal data, the datasets were anonymized, i.e. no names or addresses of TB members are indicated here, and the demographic data contained cannot be linked to specific individuals. There is also no information about the location or name of a particular TB from which the datasets come from. Some demographic data is contained, as well as data about TC exchange, a list of attributes of users and services exchanged, and a breakdown of services by date. The datasets also cover different periods of time of activity of the respective TBs. The history of Time Bank 1 was recorded in data from September 2006 to May 2014, and Time Bank 2 from October 2008 to February 2015. This does not pose a problem for individual analysis, although the TBs are directly comparable only in the period 2009 – 2013. The reason for this is that the datasets begin with the dates the respective TBs began using Time Online software, and they end when they switched to a new type of software called Time & Talents from hOurworld (hOurworld.org, 2016b). Another known factor is the size of the member bases of both TBs, which are almost the same (Time Bank 2 has 7 more members), so the TBs should be fairly comparable.

The data sets have certain limitations which make reliable analyses precarious.

- Data concerning most of the attributes (such as age, gender, nationality, religion and other such information) are not obligatory for TB members to disclose when registering. So even though some general information is contained about these features, it is not possible to relate it to particular individuals, thus it is also impossible to reliably analyze anything connected to these attributes, because there is no knowing how large a part of population answered the questions, or if all who answered some questions also answered other ones, or from which time span the answers came.
- The software Time Online was used in different periods of time, as indicated above, so data are comparable only in certain periods of time.
- Each TB uses different names for services, so it is difficult to compare specific services without knowing what exact work the name actually represents. As an example, Service DIY can exist in one TB and Hard DIY, Simple DIY, etc. in another one. In other words, there is no way of knowing which name of a service covers which activities.
- The datasets also contained some extreme values, i.e. as services which were exchanged for only 1 or 2 hours in the whole history of the data. These are most likely "dead ends," i.e. a service someone used once and then never again, and in any analysis these would create outlying values which would distort results.

These complications and dataset characteristics actually made any reliable statistical analysis impossible. When data are only gathered exclusively from a TB software unit, it was impossible to organize them into analyzable material, because there is always a part of the picture missing; terminology differs, a given time span does not cover the same timeframe; there are many outlying values, which always means that the distribution cannot be considered to be standard, etc. In order to conduct a reliable statistical analysis, this feature should already be implemented in the software, or at least, a record should be kept in an orderly manner which would make results comparable. At this point, only qualitative kinds of analyses were possible.

Systems Approach

As a part of systems thinking and systems sciences, this book searches for answers about the nature of TB as a system. In order to form a precise starting point, theoretical background entailing the classics of the General Systems Theory was researched, including the approaches of economist Kenneth Boulding, mathematicians Anatol Rapoport and Norbert Wiener, and biologist Ludvig Von Bertalanffy, as well as that of other theorists (Boulding, 2004, Rapoport, 1967, Skyttner, 2001, Skyttner, 2005, Wiener, 1951, Bertalanffy, 1969). TB is characterized by a strong practical orientation and thus a theoretical background had to be identified to progress further towards systems analysis and systems engineering.

The previously mentioned methods introduced the possibility of using systems thinking in relation to systems engineering. This also involved other systemic approaches, such as systems archetypes (Valek, 2016b, Kim and Lannon, 1997) in Time Banking, but more importantly, making an analysis and finding an appropriate methodology to break down TB as a system.

System Archetypes are present all around us as a kind of “stories” (Kim, 2000), which happen repeatedly in various environments in countless modifications. These marks of system behavior may well lead to uncovering elementary flaws in a system. The identification of System Archetypes is first step in creating a system change (Kim & Lannon, 1997). In the literature, we can find eight system archetypes (Kim, 2000), sometimes with various updates and modifications (Senge, 2006). Although systems archetypes represent a very interesting approach with a huge potential to TB, they did not prove to be the right direction for this book. They are definitely worth exploring in the future because of their practical implications in world of Time Banking, but on the level of systems design only insufficient data exist at the moment. Moreover, systems archetypes deal with systems dynamics rather than the system as a whole, and the aim of this book is general static systems design.

Notwithstanding the abovementioned overview, for this book the most valuable analytical framework proved to be Peter Checkland's Soft System Methodology.

According to the Kenneth Boulding, the 8th level of the hierarchy are the social systems which supposed to contain all human life with its complexity and richness (Boulding, 2004).

If there is a key model on which society stands which is overlooked in analyses of the market economy yet still works, we can state that there is high chance that without this important part the whole would be unsustainable in its non-systemic approach where it leaves a part of itself aside. Ignoring this vital part of the whole has yielded skewered and prejudiced analytical results and is in fact an unsustainable approach.

To implement a change in system it would be wise to use an appropriate tool. Because the economic reality around us can be considered a system, a valid approach should be a systems-oriented one. The founders of General Systems Theory, e.g. Kenneth Boulding, a Nobel prize nominee in both peace and economics, as well as biologist Ludwig Von Bertalanffy were aware of the complexity of world around us and the fact that processes leading from small simple systems to transcendental ones are interconnected. It is thus necessary to apply a holistic view. Each element, each entity which may be defined as a system is somehow related to the other systems around it. As for a clear definition, according to Derek Hitchins a system is: "A collection of interrelated entities such that both the collection and the interrelationships together reduce local entropy" (Hitchins, 1992). To clarify this along the lines of the term "Systems approach," this delineation is based on the principle that all aspects of a human problem should be treated together in a rational manner combining theory, empiricism and pragmatics (Skyttner, 2001). A system therefore is a set of elements which has the following features (Skyttner, 2001):

- Interrelationship and interdependence of objects and their attributes
- Holism
- Goal seeking
- Transformation process
- Inputs and outputs
- Entropy
- Regulation
- Hierarchy
- Differentiation

- Equifinality and multifinality

Because TB possesses all of these features it can be considered a system.

Systems Engineering

Peter Checkland outlines systems engineering as being comprised of set of activities which together lead to creation of a complex man-made entity and/or the procedures and information flows associated with its operation (Checkland, 1993). Derek Hitchins argues that there are five main reasons for systems engineering to evolve (Hitchins, 2007):

1. It has been recognized that objects which are interconnected should be perceived as a whole.
2. Systems composed of more interconnected parts can be created in a way which would make them work better and more efficient, if perceived as a whole.
3. It is evident that engineering approaches designed to control technological systems did not work when human factor is involved or system is composed of people.
4. When systems become bigger, more complex and more intricate with a higher variability of parts, designers of the system encounter problems grasping the increasingly complicated system and to predict the behavior in reality of particular elements.
5. It is evident that systems which are designed based on a precise synthesis would be able to show emergent properties along with behaviors and features which would allow designers an edge in understanding them, i.e. a comprehension greater than a mere inventory of the parts involved, one that leads to a deeper grasp of the whole in terms of properties and efficiency.

When we consider TB as a system, the need for analysis proceeds from the previous five points. TB is composed of interconnected elements which operate only as a whole. These expansive and complex elements are also entirely dependent on the human factor and thus they need to be clearly described. At the same time, emergence, i.e. the unanticipated outputs and spinoff systems that come forth as the result of a particular combination of personnel and resources through co-production, is one of key features related to the aforementioned instrumentalization of the TB concept.

In order to tackle a complex soft system, an appropriate method needs to be selected. A simple yet functional method in the world of systems engineering is the so-called V-model (Moreno, Pavon, & Rosete, 2009). This approach has been used mainly in the field of software development, with countless modifications having been adapted to meet particular needs. The V-model consists of two main parts, much like the "wings" of the letter "V." From the top of the first wing, the "project/problem definition" focuses "down" through the analysis of a problem, proceeding from the initial concept of the relevant system, through its requirements, to a detailed design, leading to the "base" of the "V," the implementation of the designed system. From there the focus is "up" the second wing, where the integration, verification and validation of a system should finally result in the successful operation of the system and its maintenance. This model is useful when making changes in hard systems in cases when the changes to be implemented are clearly defined. Although an adaptation of the V-model for soft systems has been implemented before, one extra step has been added, or to be more precise, changes have been made in the first step. Our adaptation does not begin with a particular hard system, but with a "problematic situation," in essence a much softer term. To offer even small-scale solutions toward systemic changes in many types of human-related systems, e.g. Economic, Community, Social, Educational, Healthcare, etc., TB has greatly extended the possibilities of the V-model at each stage of the approach.

Hard and Soft Systems

At this point we come to the dichotomy of a "hard" and "soft" system. These terms have to be clarified further before we progress to issues which are more closely related to TB. In his work "Skeleton of Science" Kenneth Boulding proposed the Hierarchy of Systems Complexity consisting of 9 levels (Panarchy, 2017, Boulding, 2004):

1. **Frameworks:** The geography and anatomy of the universe: the patterns of electrons around a nucleus, the pattern of atoms in a molecular formula, the arrangement of atoms in a crystal, the anatomy of the gene, the mapping of the earth, etc.
2. **Clockworks:** The solar system or simple machines such as the lever and the pulley; even quite complicated machines like steam engines and dynamos fall mostly under this category.

3. **Thermostats:** Control Mechanisms or Cybernetic Systems: the system will move to the maintenance of any given equilibrium, within limits.
4. **Cells:** Open systems or self-maintaining structures. This is the level at which life begins to differentiate itself from not life.
5. **Plants:** The outstanding characteristics of these systems (studied by the botanists) are first, a division of labor with differentiated and mutually dependent parts (roots, leaves, seeds, etc.), and second, a sharp differentiation between the genotype and the phenotype, associated with the phenomenon of equifinal or "blueprinted" growth.
6. **Animals:** Level characterized by increased mobility, teleological behavior and self-awareness, with the development of specialized information receptors (eyes, ears, etc.) leading to an enormous increase in the intake of information.
7. **Human Beings:** In addition to all, or nearly all, of the characteristics of animal systems man possesses self-consciousness, which is something different from mere awareness.
8. **Social Organizations:** The unit within a system is perhaps not the person but the "role," i.e. that part of the individual which is concerned with a particular organization or situation. Social organizations might be defined as a set of roles tied together with channels of communication.
9. **Transcendental Systems:** The ultimate, absolute and inescapably unknowable level which also exhibits systematic structures and relationships.

As seen above, the hierarchy places "Social Organizations" at the second highest level of complexity, with the following step being only transcendental systems which are above our level of comprehension. The complexity of social systems requires a more specific approach than the facile way of solving problems in so-called "hard" systems.

Hard Systems

Hard systems are designed based on a knowledge of the behavior of their parts; they are predictable and rather mechanistic. By way of typical hard systems, we can imagine, for example, construction projects, computer program design, automotive industry projects, etc. A system can be designed by the sequence mentioned in the previous section on systems engineering, and although some emergent behaviors may exist, after feedbacking and re-iteration the design should become stable and work as intended. In the case of hard systems there may be multiple points of view from the engineers involved, and it is a task of the systems engineer to adhere to the most systemic approach (Bureš, 2011). Divergent perspectives need not drastically change the essence of the whole system.

Two examples of hard system approaches will be described. The first is that of the RAND Corporation of America, created to study the interaction between science, technology and society (Skyttner, 2005). This approach has four steps:

- Problem analysis
- Generation of alternative solutions
- Evaluation of alternatives
- Selection of the optimal alternative

This strategy is quite straightforward and accesses the apparent features of a system in relation to its elements, which in hard systems (Skyttner, 2005) can be easily identified.

Skyttner (Skyttner, 2005) proposes another approach based on his self-developed DYNAMO (from DYNAmic MOdelling) modelling computer language, which consists of:

- Selecting the perspective
- Identifying system processes
- Converting it to the DYNAMO model
- Running the DYNAMO model

Soft Systems

Soft systems are not as well-structured and predictable as their hard counterparts. They are based on vague definitions and take into account human behavior (Checkland, 1999). As outlined above, systems which include the human factor are high in complexity (Boulding, 2004) and the perception of a problematic issue and design also depends on the particular points of view of the individual involved (Bureš, 2011). In this paradigm, not only systems engineers, but others who are parts/elements/objects of the system itself must participate. Soft system design requires a specific approach which would take into consideration these specifics and would be able to work them out.

The design of a soft system has no predetermined solution; it is fluid, and the solutions found are hard to test by any experiment. In short, there is no ideal best solution; there will always be only suitable ones. Although this might sound grim, without any approach to design and problem solving in soft systems the situation would remain within its original level of entropy. Thus a structured approach or methodology should be used whether we would merely like to describe and define the situation, or if the goal is to change it, improve it or solve some structure-based problem (Checkland, 1993).

The difference between the implementation of change within a hard and soft system can be illustrated as shown in Table 1.

Time Banking as a Soft System

As explained above, TB is a system. It is composed of various objects with specific attributes; it is holistic, has a goal, transforms inputs to outputs and attains its level of entropy. On the other hand, it has regulative rules and hierarchy, is differentiated. Results may be equifinal, but also multifinal. Moreover, TB is a soft system with its all various approaches stemming from different points of view based solely on people which start it, run it and empower it. This dependence on the human factor already shows that it should be considered a soft system.

TB can be defined as a soft system according to several analytical frameworks, e.g. Jordan's systems taxonomy, which classifies systems according to three properties (Jordan, 1968):

Table 1. Difference between hard and soft system

Soft System	Hard System (According to RAND)
1. Begins with an urge to bring improvement in a social system in which there is felt to be an ill-defined problem situation 2. Express the desired goal by examining elements of structure and processing and their mutual relationships. Tentative definition of system relevant to improving the problem situation 3. Formulate root definitions of relevant systems and build conceptual models of those systems 4. Improve the conceptual models using the formal system model and other systems thinking 5. Compare the conceptual models with "what is" in the real situation, and use the comparison to define desirable, feasible changes in the real world 6. Implement the agreed changes	1. Begins with an urge to solve a relatively well-defined problem which the analyst may, to a large extent, take as "given", once a client requiring help is identified 2. Analysis by examining the decision-makers' objectives as expressed in the stated need for the required system with a specified performance 3. Identify alternative systems for meeting the defined need and comparing the systems by modelling using the performance criteria 4. Select the alternative which best meets the need and it is feasible 5. No equivalent in this approach

Source: (Checkland, 1993).

- **Rate of Change:** Structural (static) or Functional (dynamic)
- **Purpose:** Purposive or Non-Purposive
- **Connectivity:** Mechanistic or Organismic

According to these properties, TB is a functional, purposive and organismic system, which places it in terms of complexity on the same level as living organisms.

Progressing from this to Checkland's "map of systems," we can classify TB as a human activity system in which the origin is man's self-consciousness (Checkland, 1993). This finally brings us to final step of defining TB as a "social system," whereby social systems are soft in nature, as they are groupings of people who are aware of and acknowledge their membership in the group. They accept various responsibilities as a result of their membership and have certain expectations of other members (Checkland, 1993). To sum it up, TB is a functional, purposive, organismic social system.

REFERENCES

Anthes, G. (2016). Open Source Software No Longer Optional. *Communications of the ACM*, *59*(8), 15–17. doi:10.1145/2949684

Beamer, L. (1992). Learning Intercultural Communication Competence. *International Journal of Business Communication*, *29*(3), 285–303. doi:10.1177/002194369202900306

Bertalanffy, L. V. (1969). *General system theory; foundations, development, applications*. New York: G. Braziller.

Blanc, J. (2011). Classifying "CCs": Community, complementary and local currencies' types and generations. *International Journal of Community Currency Research*, *15*(D), 4–10.

Boulding, K. (2004). General systems theory - the skeleton of science. *E:CO, 6*, 127-139.

Boyle, D. (2014). *The Potential of Time Banks to support Social Inclusion and Employability. JCR Scientific and Policy Reports, Report EUR 26346 EN*. European Commission.

Boyle, D., & Bird, S. (2014). *Give and Take: How timebanking is transforming healthcare*. Timebanking UK.

Bureš, V. (2011). *Systemove mysleni pro manazery*. Pribram, Czech Republic: Professional publishing.

Bureš, V., & Tučník, P. (2014). Complex Agent-based Models: Application of a constructivism in the Economic Research. *E+M. Ekonomie a Management, 17*(3), 152–168. doi:10.15240/tul/001/2014-3-012

Cahn, E. S. (2000). *No more throw-away people: the co-production imperative*. Washington, DC: Essential.

Cahn, E. S. (2001). On LETS and Time Dollars. *International Journal of Community Currency Research, 5*, 1–4.

Cahn, E. S., & Rowe, J. (1998). *Time dollars: the new currency that enables Americans to turn their hidden resource--time--into personal security & community renewal*. Chicago: Family Resource Coalition of America.

CCRC. (2016). *Complementary Currency Resource Center*. Available: http://complementarycurrency.org/

Checkland, P. (1993). *Systems Thinking, Systems Practice*. Chichester, UK: John Wiley & Sons Ltd.

Checkland, P. (1999). Soft systems methodology: A thirty year retrospective. *Systems Research and Behavioral Science, 17*(S1), 11–S58. doi:10.1002/1099-1743(200011)17:1+<::AID-SRES374>3.0.CO;2-O

Clement, N., Holbrook, A., Forster, D., Macneil, J., Smith, M., Lyons, K., & Donald, E. M. (2017). Timebanking, Co-production and Normative Principles: Putting Normative Principles into Practice. *International Journal of Community Currency Research, 21*, 37–52.

Collom, E., Lasker, J. N. & Kyriacou, C. (2012). *Equal Time, Equal Value: Community Currencies and Time Banking in the US. Equal Time*. Academic Press.

Demeulenaere, S. (2008). Annual report of the worldwide database of complementary currency systems. *International Journal of Community Currency Research, 12*, 2–19.

Fiske, A. P. (1992). The 4 elementary forms of sociality - Framework for a unified theory of social-relations. *Psychological Review, 99*(4), 689–723. doi:10.1037/0033-295X.99.4.689 PMID:1454904

Gregory, L. (2012). Time and punishment: A comparison of UK and US time bank use in criminal justice systems. *Journal of Comparative Social Welfare*, *28*(3), 195–208. doi:10.1080/17486831.2012.742452

Hale, S., Leggett, W., & Martell, L. (2004). *The Third Way and Beyond: Criticism, futures and alternatives, Oxford Road.* Manchester, UK: Manchester University Press. doi:10.7228/manchester/9780719065989

Hallsmith, G., Lietaer, B. A., Lovins, L. H., & Meadows, D. L. (2011). *Creating wealth: Growing local economies with local currencies.* New Society Publishers.

Hildreth, P., Kimble, C., & Wright, P. (2000). Communities of practice in the distributed international environment. *Journal of Knowledge Management*, *4*(1), 27–38. doi:10.1108/13673270010315920

Hitchins, D. K. (1992). *Putting systems to work.* Chichester, UK: Wiley.

Hitchins, D. K. (2007). *Systems engineering: a 21st century systems methodology*. Chichester, UK: John Wiley. doi:10.1002/9780470518762

Holmgren, D. (2011). *Permaculture: Principles & Pathways Beyond Sustainability*. Permanent Publications Hyden House Ltd.

hourworld.org. (2016). *Main Page*. Available: hourworld.org

IJCCR. (2017). Available: https://ijccr.net/

Ilin, E. P. (2013). *Psikhologia pomochi: Altruism, Egoism, Empatia*. Sankt-Peterburg: Piter.

Jevons, W. S. (1875). *Money and the mechanism of exchange*. London: H.S. King & Co.

Jordan, N. (1968). Themes in speculative psychology. London: Tavistock Publications.

Kennedy, M., Lietaer, B., & Rogers, J. (2012). *People money: The promise of regional currencies*. Axminster, UK: Triarchy Press.

Kim, D. H. (2000). Systems Archetypes I: Diagnosing Systemic Issues and Designing High-Leverage Interventions. Waltham, MA: Pegasus Communications, Inc.

Kim, D. H., & Lannon, C. P. (1997). *Applying Systems Archetypes*. Pegasus Communications, Inc.

Lehaney, B. (2004). Beyond Knowledge Management. Idea Group Publishing.

Lesser, E. L., & Storck, J. (2001). Communities of practice and organizational performance. *IBM Systems Journal*, 40.

Lietaer, Arnsperger, Goerner, & Brunnhuber. (2012). *Money and sustainability: the missing link*. Club of Rome-EU Chapter to Finance Watch and the World Business Academy.

Lietaer, B. (2001). *The Future of Money*. London: Century, Random House Group Limited.

Lietaer, B. A., & Dunne, J. (2013). *Rethinking money: how new currencies turn scarcity into prosperity*. San Francisco: Berrett-Koehler.

Linderman, M., Siegel, B., Ouellet, D., Brichacek, J., Haines, S., Chase, G. & O'May, J. (2005). *A Reference Model for Information Management to Support Coalition Information Sharing Needs*. Academic Press.

Maes, R. (1999). *Reconsidering Information Mangagement Through A Generic Framework*. Amsterdam: Universiteit van Amsterdam.

Maes, R., Rijsenbrij, D., Truijens, O., & Goedvolk, H. (2000). *Redefining business - IT alingment through a unified frameworks*. Universiteit van Amsterdam.

Manville, B., & Foote, N. (1996). Harvest your workers´ knowledge. *Datamation*, 7.

Martignoni, J. (2012). A new approach to a typology of complementary currencies. *International Journal of Community Currency Research, 16*, A1–A17.

Mill, J. S. (1900). *Principles of Political Economy*. London: Longmans, Green, and Co.

Naughton-Doe, R. (2015). *An evaluation of timebanking in England: What can timebanks contribute to the co-production of preventive social care?* (Dissertation). University of Bristol.

Nonaka, I. (1991). *The Knowledge-Creating Company*. Harvard Business Review.

Nonaka, I., & Takeuchi, H. (1995). *The Knowledge-Creating Company*. Oxford, UK: Oxford University Press.

Otcenaskova, T., Bures, V., & Mikulecka, J. (2012). Theoretical Fundaments of Knowledge Intensity Modelling. In *18th International Business Information Management Association conference*. Istanbul, Turkey: IBIMA.

Panarchy. (2017). *Boulding's Classification of Systems*. Available: https://www.panarchy.org/boulding/systems.1956.html

Papaoikonomou, E., & Valor, C. (2016). Exploring commitment in peer-to-peer exchanges: The case of timebanks. *Journal of Marketing Management*, *32*(13-14), 1333–1358. doi:10.1080/0267257X.2016.1177578

Papaoikonomou, E., & Valor, C. (2017). The Institutionalization of Alternative Economies: The Processes of Objectification in Time Banks. *Journal of Macromarketing*, *37*(2), 167–179. doi:10.1177/0276146716672286

Rapoport, A. (1967). *Games which simulate deterrence and disarmament*. Clarkson, Ont.: Canadian Peace Research Institute.

Samuelson, P. A., & Nordhaus, W. D. (2005). *Economics*. Boston: McGraw-Hill/Irwin.

Sandu, A. (2012). *Appreciative Ethics*. Iasi, Romania: Lumen publishing house.

Schroeder, R., Miyazaki, Y., & Fare, M. (2011). Community Currency Research: An analysis of the literature. *International Journal of Community Currency Research, 15*(A), 31-41.

Senge, P. (2006). The Fifth Discipline: The Art and Practice of the Learning Organization. New York: Currency/Doubleday a division of Random House, Inc.

Seyfang, G. (1999). The Euro, The Pound, And The Shell In Our Pockets: Rationales for Complementary Currencies In A Global Economy. *New Political Economy*, *5*(2), 227–246. doi:10.1080/713687774

Seyfang, G. (2006). Sustainable consumption, the new economics and community currencies: Developing new institutions for environmental governance. *Regional Studies*, *40*(7), 781–791. doi:10.1080/00343400600959173

Seyfang, G., & Longhurst, N. (2012a). Community Currencies and Sustainability. In U. O. E. Anglia (Ed.), *Grassroots Innovations Research Briefing*.

Seyfang, G., & Longhurst, N. (2012b). *Money, money, money? A scoping study of grassroots complementary currencies for sustainability*. Norwich: Science, Society and Sustainability Research Group.

Shreiber, G., Akkermans, H., Anjewierden, A., Hoog, R. D., Shadbolt, N., Velde, W. V. D., & Wielinga, B. (2002). Knowledge Engineering and Management. Cambridge, MA: The MIT Press.

Skyttner, L. (2001). *General systems theory: Ideas & applications*. River Edge, NJ: World Scientific. doi:10.1142/4307

Skyttner, L. (2005). *General systems theory: Problems, perspectives, practice*. Hackensack, NJ: World Scientific.

Smutek, M. (2017). Life-world and System in Social Services. In Hradec Days of Social Work: Social Work as a Means of Promoting Social Change, Hradec Kralove, Czech Republic.

Svaricek, R., & Sedova, K. (2014). *Kvalitativni vyzkum v pedagogickych vedach [Qualitative reseach in pedagogic sciences]*. Prague, Czech Republic: Portal.

TBUSA. (2016). *About Timebanking USA*. Available: http://timebanks.org/about/

Tesfatsion, L. (2003). Agent-based computational economics: Modeling economies as complex adaptive systems. *Information Sciences*, *149*(4), 263–269. doi:10.1016/S0020-0255(02)00280-3

The Economist. (2013, November 30). Digital money: The bitcoin bubble. *The Economist*.

Timebanking UK. (2005). A bridge to tomorrow: time banking for baby boomers. Gloucester, UK: Timebanks UK.

Timebanking UK. (2015). *Time Banking types*. Available: http://www.timebanking.org/what-is-timebanking/what-is-timebanking/

Tucnik, P., Valek, L., Blecha, P., & Bures, V. (2016). Use of Time Banking as a non-monetary component in agent-based computational economics models. *WSEAS Transactions on Business and Economics, 13*.

Valek, L. (2013a). Risk Factors and Issues Which Might Arise While Implementing and Running a Time Bank: Learning from Success and Failure. In *22nd International-Business-Information-Management-Association Conference on Creating Global Competitive Economies: 2020 Vision Planning and Implementation*. Rome, Italy: Int Business Information Management Assoc-Ibima.

Valek, L. (2013b). *Time Banks in Czech Republic: Filling an Empty Gap in Time Bank Research*. Int Business Information Management Assoc-Ibima.

Valek, L. (2014). The Time Bank implementation and governance: Is PRINCE 2 suitable? In *Centeris 2014 - Conference on Enterprise Information Systems / Projman 2014 - International Conference on Project Management / Hcist 2014 - International Conference on Health and Social Care Information Systems and Technologies*. Amsterdam: Elsevier Science Bv.

Valek, L. (2015). The difference in understanding of time banking in various contexts. In *6th LUMEN International Conference: Rethinking Social Action. Core Values 2015*. Iasi, Romania: LUMEN.

Valek, L. (2016a). Open Ways for Time Banking Research: Project Management and Beyond. *International Journal of Human Capital and Information Technology Professionals*, *7*(1), 35–47. doi:10.4018/IJHCITP.2016010103

Valek, L. (2016b). Time Banking and its place in economic education. *Journal of Economics and Economic Education Research*, *17*, 9–15.

Valek, L., Kolerova, K., & Otcenaskova, T. (2014). Time banks and clusters: Similarities of sharing framework. *Global Journal on Technology*, *6*, 31-36.

Valor, C., & Papaoikonomou, E. (2016). Time Banking in Spain: Exploring their Structure, Management and Users' Profile. *Revista Internacional de Sociologia*, *74*(1), 14. doi:10.3989/ris.2016.74.1.028

White, L. H. (2015). The merits and feasibility of returning to a commodity standard. *Journal of Financial Stability*, *17*, 59–64. doi:10.1016/j.jfs.2015.03.003

Wiener, N. (1951). *Cybernetics and society*. New York: Executive Techniques.

Chapter 3
Systemic Analysis of TB Concept

ABSTRACT

The previous chapter contains the introduction of main applied methods and approaches. This chapter provides the description of the soft system methodology and explanation of single steps needed to be applied in practice. This method is employed in a reduced PQR form. Root definitions and unstructured problem descriptions help to depict the concept of TB. Moreover, charts and diagrams enabling the provision of more accurate insight, and the identification of main inputs, entities, and processes of the system are developed. Finally, entities that belong to the wider system are delineated. In general, this chapter points out the lack of scientific background within the current research. A novel TB taxonomy as an appropriate starting point aiming at the solution of this issue is proposed. Due to the deconstruction of TB to the particular main elements and processes, the soft system methodology helps to uncover the real core of TB.

INTRODUCTION

At this point it is necessary to select a methodology which would allow us to tackle unstructured and vague questions such as "Is there an universal approach to TB?" or "How to ensure implementation of TB concept in various fields securing its benefits?" Peter Checkland's Soft System Methodology (SSM) (Checkland, 1993) seems to be perfectly suitable for this. As is apparent from

DOI: 10.4018/978-1-5225-6974-9.ch003

previous chapters, we will not "design" a system and we are past the stage of engineering, because TB already exists. The purpose of applying SSM is to search in various applications of the TB concept for common features and the main building elements, to identify and describe them, and to draw conclusions. Only then would the way be open for drafting a universal TB system. In other words, by using SSM we search for an answer to the question of what TB is (a unified definition) based on what we know in terms of what a system does in most situations.

Since one of main advantages of SSM is the possibility of generating a completely new point of view, i.e. not merely superficially tweaking already existing systems, it can be said that the TB concept is projected to bring about a systemic change. Nevertheless, with the anticipated power of TB as a tool to clearly define a specific systemic change, fundamental questions remain to be structured such as how to implement it correctly in various conditions and how to anticipate possible consequences. In other words, in combining this relatively new system of TB with SSM to prescribe and describe a change, we fully expect surprising positive and negative side effects, with managing these a natural and anticipated aspect of this holistic process (Valek, 2016).

The Soft System Methodology

The SSM of Peter Checkland is divided into seven main steps/stages. Stages 1, 2, 5, 6 and 7 belong to the "real world," with 1 and 2 extracting information from the problem situation and 5, 6, 7 comparing and drawing possible solutions for a change. Stages 3 and 4 belong to the "systems thinking layer" where root definitions are defined and transformed to conceptual models (Checkland, 1993).

Unstructured Problem Situation and the Problem Situation Expressed

At this point it is necessary to create a picture of the problematic situation, not only a problem (Checkland, 1993). From this point of view, the problematic situation would be the implementation of TB itself. The word "problematic" might be confusing, as it is often perceived as something negative (Checkland, 1993). At this point the term is used only in order to describe a situation which has to be solved. This situation should be expressed in richest possible

Figure 1. Structure of the Soft System Methodology
Source: (Checkland, 1993).

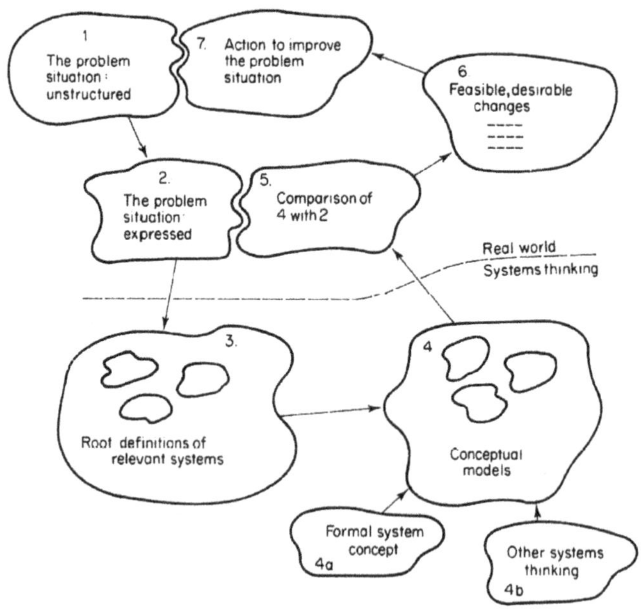

way (also with help of a method called "rich pictures"). At the second stage, the structure of the problematic situation must be clarified, possibly even describing sub-problems within the situation; processes which are present within these layers should also be uncovered. The structure might be examined in terms of physical layout, power hierarchy, reporting structure, and the patter of communications both formal and informal. (Checkland, 1993). Processes may be examined in terms of the basic activities of deciding to do something and doing it, along with monitoring both of these steps. Processes may be also evaluated in terms of their external effects and corrective action(s) to be taken. The relationship between structure and process has frequently been found to be a core characteristic of examining situations in which problems are expected (Checkland, 1993).

Practically for this book, all this means that an unstructured problematic situation has to be identified and expressed. This will be done in detail in the chapter featuring the analysis of a TB situation, which will include the formulation of a completely new Taxonomy of TB Approaches as an analysis of the problematic situation. As the problematic situation is expressed, it will be necessary to create models of the approaches to TB described in the Taxonomy as well as to describe them in detail.

Root Definitions of Relevant Systems

With a focus on what the systems (elements of problematic situation) are, not on what they do (Checkland, 1993), these definitions must be broad as well as include possible restrictions in order to count on them for further use. It is desirable to create quality definitions which would allow us to produce suitable conceptual models. To assure the connection between these two factors and reality, it is recommended to use the so-called CATWOE model (Checkland, 1993). The acronym proceeds from the following: C – customers: the entity for which the change is done and who are affected by it. A – actors: the agents who carry out main activities of the system, especially its transformation. T – transformation: the means by which defined inputs are transformed into defined outputs. W – Weltanschauung: an outlook, framework or image which makes this particular root definition meaningful. O – ownership: an entity which have prime concern over the system and can cause the system to cease to exist. E – environmental constraints: features of system's environment which must be taken as given (Checkland, 1993).

In our context CATWOE can be defined as follows:

- **Customers:** All entities influenced by TB operation. This includes not only members, but also people and organizations outside of the system, i.e. potential stakeholders, partner organizations as well as individuals influenced by spin-off projects and activities.
- **Actors:** TB members as well as other entities within the TB system.
- **Transformation:** A transformation process happens within a TB system and is facilitated by entities which are grouped into various sub-systems.
- **Weltanschauung:** The analytical point of view regarding sub-systems within a system, with the possibility of breaking down the system to an additional level of sub-sub-systems within a sub-system. Any deeper resolution, however, would cause redundant complexity.
- **Ownership:** This depends on the particular point of view, but in essence ownership is categorized as a condition of any entity which is either the founder of a base organization or which has a strong influence on a system, such as government with a system of laws, etc.

- **Environment:** Usually a region where is a problem to solve and a simple tool is needed to not only solve the problem, but to also exert a strong positive impact on the economy and society there.

Root definitions will be produced as a result of a rich picture and element analysis of the available approaches as indicated above. They will serve to summarize the process of modelling the outlined approaches.

Conceptual Models

Building conceptual models on the basis of root definitions is the most rigorous step in the whole process of SSM. Root definitions describe a system as what it is. A conceptual model should describe what it should do in order to become what is described in the definition. A key point is a description based on verbs in logical order which would lead to the transformation. It is recommended to start with about six verbs which cover the main activities from root definitions. We also might describe sub-systems as a higher "resolution level," or these sub-systems might emerge as a part of the change that follows. At this point when defining a system, particular features of what can formally be called a system must be considered. A conceptual model must have an ongoing purpose or mission and a measure of performance; it contains a decision-taking process and contains components which are themselves systems encompassing all properties of a system. A conceptual model has components which interact, exists within wider systems and/or environments with which it interacts, but has a boundary separating it from them. A conceptual model contains resources, both physical and abstract through its human participants. It entails some guarantee of continuity, has long-term stability, and recovers stability after some degree of disturbance (Checkland, 1993).

In this book, based on two previous steps, each of the elements from all approaches will be isolated and a new universal system will be proposed and described.

Comparison of the Situation With Developed Models

This stage could in practice employ various approaches, but here four main ones will be employed. The first approach is to use conceptual models to devise questions related to the existing situation. When answered, these inquiries should shed light on differences between the model and reality. The second approach is to reconstruct events from the past and compare what would happen if conceptual model would have been used at various points in the organization's history. The third approach is to discuss the conceptual model with an appropriate stakeholder who can identify the main differences. And finally, the fourth approach is based on the creation of a model of what now exists and then comparing this model of reality with the conceptual one.

In the last section of this book a comparison will be made with the original approaches. Conclusions regarding the new universal system in terms of its relationship with the fields of knowledge and information management will be made along with projections regarding the new system's use in a scenario involving a potential IT solutions project.

Feasible Desirable Changes and Action to Improve the Problem Situation

The changes can now be outlined and proposed based on the previous comparison. There can be changes of 3 kinds: in structure, in procedures and in attitude. As the first two are relatively easy to achieve, changing attitudes might be very difficult. In addition, it has to be noted that these changes must be systematically desirable as a result of root definitions as well as culturally feasible given the characteristics of the situation and people involved (Checkland, 1993).

These two last stages are not applicable in this case and are introduced only to show SSM in its complete form. This book is operating on a conceptual level and the proposal of universal system with broader context is already entailed in the goal.

Application of SSM on TBs

Even though the SSM is an universal tool for tackling soft systems (Checkland, 1993) and the steps proposed above will be tailored to this book, according to revisions (Checkland, 1999) minor changes have been proposed, especially in the design of rich pictures. This comes from the fact that this book focuses on a conceptual point of view, and as such it will uncover basic approaches to TB to create one universal approach. It has been already demonstrated that SSM is appropriate method, but at this premature stage it is not feasible to conduct this type of analysis to the fullest extent possible. An analysis of the basic components of the TB system might appear similar to a hard systems analysis because of its expected schematic features, but these elements are necessary in order to progress further to a synthesis. We would obfuscate the issues entailed in, e.g. gaining a structured picture of the main elements, relations and flows, if a hard systems approach was used hen there is corroboration that TB is a soft system. SSM should be used, but in a simplified form. The first reason is the fact that to progress further to a full scale SSM we must build a solid basis using systems context of TB (an overview of the system). The second reason is more practical, i.e. that extent of this book does not allow to continue to next level, as to conduct a full scale SSM analysis involving all possible points of view would take much more space than there is in this book. Moreover, this is not entirely possible without first obtaining a general overview and determining a systems definition. However, even a simplified SSM approach is a first step towards a conceptualization of a systems view of TB and necessary one for synthetizing a universal model and definition. The simplified approach will also establish the "weltanschauung" element of SSM, which does not disappear from the focus, but is fixed as a standard to allow comparison of all the analyzed models.

Design of Rich Pictures

In order to keep a consistent point of view of the various approaches towards TB - to maintain the same weltanschauung - the structure of rich picture design will be based on three main questions which define root definitions, the next step in SSM. In this step, these definitions will have already been structured in a standardized way, a process which anticipates the next step. In other words, the weltanschauung will proceed from the root definitions, which consist of the following (Checkland, 1999):

- What to do? - P
- How to do it? - Q
- Why to do it? - R

As simple as this might sound, these are fundamental questions which define the system we are describing (P), possible sub-systems (Q), and the wider system (R) that surrounds the described one, of which P is a sub-system (Checkland, 1999). This approach keeps focus on a particular system, allowing the exploration of adjacent ones without the limitation of including the elements mentioned above. It shows how to proceed as well as which inputs to the analyzed system come from the wider system at the ownership level (entities who have a main interest in the existence of the system), and monitors outputs and feedbacks out of a system. System level analysis shows objects, connections, flows and processes within a system with relation to its sub-systems, which can be then broken down to individual elements as well. The idea of PQR is also supported by other sources which recommend asking the same questions even in other contexts (LMC, 2017, Shreiber et al., 2002).

This approach also keeps the rich pictures structured and informative enough to provide the resources of any system in one page. Other approaches exist, such as Tiger spreadsheet solutions (Tiger Spreadsheet Solutions, 2016), which offers a 6-step approach with the advantage of not being centered on the design or drawing skills of the individual who is creating the rich picture. The Tiger system can be implemented by following the structure which is proposed above: creating a list of issues for the main problematic situation; deciding on focus, i.e. W, or the weltanschaung of the CATWOE approach; focusing on relationships, which is natural in any systems approach; focusing on inputs, outputs and processes within, which is the main key to decipher what is happening.

Based on this information a completely new approach was devised. The approach of PQR created a unique frame of a circular design where inside there is a system: "P", which has its connections towards the sub-systems "Q", which are encompassed in the wider system (the "R"). This division and the structure will define the following descriptions of all nine TB systems. This approach is also unique because it proceeds from many iterations in application of SSM and makes it more simple in the way that it already includes root definitions in the previous step and prepares the output for subsequent steps. SSM, although originally a tool for systems engineers, is in this way simplified for common use.

To extend the previously mentioned argument, by this simplification and fixed "weltanschauung" we are getting closer to hard systems due to the fact that we are searching for fundamental connections in TB based only on descriptive and vague information. This is why a soft systems approach is necessary, as the problematic situation is very fluid, i.e. depending mostly on tacit knowledge and human factor. But by the same token we need to determine fixed describable connections within the situation to be able to describe it or even (re)design it. Only after this may full scale SSM be attempted, i.e. after attempting to assume the points of view of various stakeholders, users, environments, etc. On the other hand, focus is centered on transformation process (T) in area of system P, where inputs are transformed to outputs by interaction of internal entities (subsystems) Q. This emphasis is imperative in order to study and describe the nature of the system itself.

Example of a PQR Scheme

The use of the PQR approach in this book is based on countless iterations of the application of the first four steps of the SSM. The approach which was devised simplifies the matter greatly, and the scheme of system is settled in a specific form. The main core is already implementing the perspective of CATWOE in drawing a rich picture or a diagram of a system. Most important is W – weltanschauung – where the "resolution" of the picture is selected, i.e. the main system being studied, its sub-systems as well as outside influences. To keep this resolution working, it is necessary to maintain a certain level of discipline and not penetrate too deeply into the various subsystems, in which case the resolution would change, and a single subsystem may reveal in too much superfluous detail its own subsystems and their subsystems. For this reason, W is fixed (as explained above) and the originally selected W should be strictly adhered to until a picture of desired system is clearly drawn, and then we would proceed to a deeper exploration of its subsystems. Applying CATWOE in the stage of expressed problematic situation T is considered as a space of the system where the transformation of inputs to outputs occurs, a step which is in the PQR model designated as P, with this transformation also using subsystems Q. This is all encompassed in the wider system R, which contains O and E from CATWOE. Thus, the level of ownership has control over the system, but also level of external environment. Finally, on all levels we have to consider C and A of CATWOE – Customers and Actors, i.e. those for whom the transformation is being performed and who are the entities allowing the transformation to happen. The result is a circular structure in which the inner circle contains P with its Qs, and the outside of the circle is the domain of R. An example related to the TB system is shown in the diagram below.

This book brings an innovative point of view of TB by using the SSM approach, but it also uses the steps of SSM in its application section. In order to learn more about TB as a system, it is necessary to break it down into the main components, to identify them and to find a structure. Firstly, the main

Figure 2. Example of circular model for TB environment
Source: (author's research).

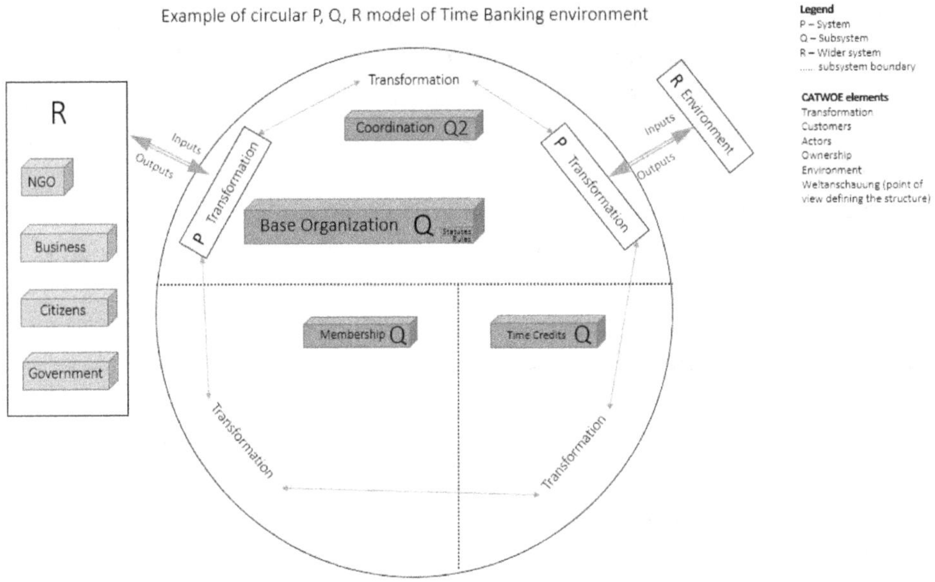

approaches to TB have to be defined and the problematic situation needs to be properly described. After this step we can go deeper and draft conceptual models.

Only at this point can a decision about how to analyze the system come. Systems analysis is a mental process of learning about the world using systems thinking and a systems approach. The world is composed of various elements, objects, processes, phenomena, problems, etc. (Bureš, 2011). In case of a soft system, the situation becomes very abstract, as SSM contains its own inherent analytic tools, i.e. so-called rich pictures, which will be determined in steps 1 and 2 of SSM by this method.

Problem Situation Unstructured

The problem situation is a non-existent systemic framework for TB, including a TB taxonomy based on systems elements of TB as a soft system and a systemic definition.

This situation may in fact have more solutions, with aim of this book being to analyze Time Banking as a system, break it down into the basic elements, to determine the main relationships and bonds among them, and to re-construct a universal system which would allow the problem situation to be solved.

In order to identify the main elements of a TB as a system and describe main ones, it is necessary to summarize approaches to TB which exist. There are several TB models in use around the world and there are several typologies. Nevertheless, there is no systems based taxonomy would fit the aims of this book.

Synthetizing a systemic definition will be possible only at very end of this book, when all the features of all the approaches as well as the nature of a universal TB has been determined.

Taxonomy of Time Banks

A time bank classification or a taxonomy is another problematic area. To this date there has been no clear taxonomy of TBs. In the past there have been many attempts to catalog types of TBs and somehow classify the category (Ryan-Collins et al., 2008, Naughton-Doe, 2015, Timebank Mahoning Watershed, 2017, Blanc, 2011, Timebanking UK, 2015, Boyle and Bird, 2014). Still, the use of TB remains subjective according to the point of view of a particular person or a researcher, which is perfectly normal, since as a soft system TB can be viewed from many perspectives. To form a solid basis for this book, however, a taxonomy based on the literature and previous research will be created. The following chapter summarizes the main approaches towards TB classification undertaken so far by other researchers, thus facilitating the creation of one listing based on the main common features, which can then be isolated to cover the entire spectrum of possible additions to the taxonomy of the present book.

Known Classifications

Naughton-Doe introduces the classifications shown in Table 1.

This classification comes in coupled variants in the perspective the researcher needed for her research. This cataloguing brings important information for the further development of our taxonomy.

Another classification derived from Welsh model can be found in (Timebank Mahoning Watershed, 2017):

Table 1. Time Bank classification by Naughton-Doe

Key Terms	Definition
Top-down	Timebanks started or commissioned by an organization.
Bottom-up	Timebanks started by an individual or individuals.
Community-Based	Timebanks consisting of either separate organizations, or hosted by an organization but managed separately.
Organization-Based	Timebanks hosted by organizations and which have adopted the host organization's aims, policies and culture.
Homogeneous members	Timebanks that have recruited a specific demographic for their membership, for example, older people or people with mental health problems.
Heterogeneous Members	Timebanks that have recruited the general community for their membership.
Target driven	Timebanks that have targets, usually imposed by funders or host organizations. For example, numbers of members and numbers of exchanges.
No targets	A timebank with no formal targets.
Formal policies	A timebank that has adhered to rigid policies, rules and regulations, often imposed by a host organization.
Informal policies	A timebank able to establish its policies, rules and regulations independently and has a relaxed, flexible environment.
Activist time broker	Brokers motivated by personal values and invested personal time in the timebank.
Careerist time broker	Brokers motivated by career progression.
Employee time broker	Brokers who regard their role as a job and do not invest personal time in the timebank.

Source: (Naughton-Doe, 2015).

- **Person to Person Model (1 to 1 Exchanges):** A mutual (open) credit system whereby credit exists solely between individuals. Exchanges which take place between individuals are initially set up by a central hub. This model facilitates acts of neighborliness between individuals
- **Agency to Client Model (Specialist Time Banking):** A mutual (open) credit system, whereby credit exists between the agency/ organization and person. Activity is themed, based around the specialist skills of agency leading the time bank. Time banking embedded within an existing organization, voluntary or statutory.
- **Agency to Client Model:** A fiat (closed) credit system, with credits originating from the 'bank.' An initial time audit establishes the current level of active community involvement. The number of credits

available is based on a time audit along with projections for future involvement.

- **Organizational Time Banking:** A mutual (open) credit system where credit exists between agencies. Used to facilitate exchanges between organizations for mutual benefit and to enhance the use of scarce or underutilized resources.
- **Institutional Time Banking:** A mutual (open) credit system, where credit exists between the agency/ organization and person. This time bank is situated within an employment setting, with membership of the time bank open to employees, staff, students, etc. Activity may be themed around the particular needs of the agency leading the time bank, or may be more focused on meeting the personal needs of staff/ employees. Using a time banking tool to create learning organizations.
- **Time Centre Model:** A fiat (closed) credit system, with credits originating from the 'bank.' The time banking tool addresses problems of under capacity and ensures that those people taking part are actively involved. An initial time audit establishes the current level of active community involvement. The number of TCs available is based on a time audit and projections for future involvement.
- **Time Network Model:** A fiat (closed) credit system, with credits originating from the 'bank.' All local service providers, both voluntary and statutory, use time banking as their currency to recognize and reward community involvement.

This very different classification illustrates how the literature is divided in terms of basic attitude towards TB as well as how each particular researcher might create unique definitions based on the needs of the research. To form a systems analysis (Bureš, 2011), it would be necessary to organize these ideas into a sensible framework which would consider the point of view of systemic elements, and thus describe the system as it is, not as what it does (Checkland, 1993). To get at the core of TB as a system, it should be classified in an atomic way so that the classification can be concretized into a final mode.

A most basic division in this sense seems to be the one of Ryan-Collins (Ryan-Collins et al., 2008). Here the general concept of a person-to-person, agency-to-person and agency-to agency TB is presented, categories which have been confirmed by other authors (Ryan-Collins et al., 2008, Valor and Papaoikonomou, 2016, Molnar, 2011). These three branches seem to be elemental and the basis for other divisions and the separation facilitates the description of a wide variety of other approaches to TB. This division into these three branches is supported, with slightly different wording, by other resources such as TBUK and TBUSA (Timebanking UK, 2015, Timebanks USA, 2017). Marks speaks about Neighbor-to-neighbor and Individual-to-Institution variants (Marks, 2012), which are also basically the first two branches. Possible sub-systems of these branches may be for example management, membership concept, target group, base institution as well as others which would be identified after an examination of the last classification.

The main three branches of TB are shown in Table 2 along with descriptions of the role of the Agency/Base organization as well as key outcomes according to (Ryan-Collins et al., 2008).

The P2P approach is generally used when a TB is applied as a social support system between individuals. Individuals are part of an established TB "hub" or center, often hosted by a third sector or public-sector organization and run by a coordinator. Members exchange mutual credit in the form of hours (time credits), with a coordinator overseeing this procedure. Every unit of currency in this model is created by an agreement and an exchange; each unit is earned back by an equal obligation to 'pay back' by the beneficiary of the exchange. Members with debits are to be encouraged to do jobs and earn credits, whilst those already possessing credits are be encouraged to spend them in order to increase the circulation of currency (Ryan-Collins et al., 2008).

Table 2. Time Bank classification by Ryan-Collins

Approach	Role of Agency	Key Outcomes
a) Person-to- Person (P2P)	Host Funder of time bank broker/ administrator Individual organizational member	Builds social capital and community networks Engages hard-to-reach areas Creates a multiplier effect of giving and receiving
b) Person-to-agency (P2A)	Catalyst to engage people in the design and delivery of services Issuer of credits and creator of rewards with partners	Builds social capital and community networks Engages hard-to-reach areas Transforms service delivery through co-production Sustainable and community-led services
c) Agency- to-agency (A2A)	Collaboration Shared learning Shared resources and assets	Better services through the more efficient use of resources Greater efficiency in terms of use of agency assets More sustainable future

Source: (Ryan-Collins et al., 2008).

The P2A approach allows people earn time credits through their contribution to the aims and goals of an agency or community, whether public, third sector, or business. The agency itself is the 'bank' (Marks, 2012) and issues time credits in recognition of the time and effort put into the overall goals. The agency also provides a range of options for people to spend these credits. These typically include community-based activities, trips, services and events organized that help to further build social networks and create a sense of collective identity. These frequently employ existing underutilized resources such as sporting facilities, empty rooms or access to IT (Ryan-Collins et al., 2008).

The A2A TB approach signifies TB as a tool that can help facilitate exchanges between organizations with a common purpose as well as between individuals. There are very few organizations, including commercial organizations, that fully utilize the assets and staff capacity at their disposal. Empty meeting rooms, stationary mini-buses, 'down-time' when staff or volunteers have little to do, wasted food and energy, old IT equipment: the list is potentially endless. How many organizations consider sharing such underutilized assets (Ryan-Collins et al., 2008)?

To summarize this approach, it is not reserved exclusively for the so-called "third sector," or non-profit non-governmental organization (NGO), but it also allows also businesses to join in. It is even possible to create a network of companies which would be able to share unused resources in a similar way as do individuals in P2P TB; in fact, companies can indeed be the very basis of the A2A approach. This idea is actually not far from the business exchange model noted in the state-of-the-art chapter of the present work and which has already been in use since the 1930s (Kennedy et al., 2012), or newly forming Enterprise Ethereum Alliance (EEA, 2018). The simplicity and universality of the TB approach makes it the ideal tool not only for utilization of a company's assets which are not being used at a particular moment, but also to offer assets which might serve a different purpose for a peer organization in a TB, thus gaining additional value. These additional assets may be secondary (in relation to their work assignments) knowledge or skills of employees which are not being utilized by host company at the moment, or at all. This may also serve as a space, a set of tools or other resources which partner organizations would use for other purposes than the host would anticipate (for example use of one company's empty room as a training room for employees of another company). From the previous paragraphs it seems that the P2P approach works best for individuals, whereas the A2A approach would be used between organizations. A2A has a great potential in the business sphere, especially in the areas which allow this kind of cooperation, such as company clusters (Valek et al., 2014).

All of these features will be defined and examined later in this book, but before we progress further to underlining this interoperability of the TB concept, the TB should be described as a system, with the main elements consisting of sub-versions or sub-systems. As for the terminology described in the text above, a change is proposed. For the purposes of this book the word "agency" may be misleading, with a more appropriate term emerging as "organization," with the latter term's implication of people coordinating themselves in a structured way, whether it be through business, NGOs, governmental organization, or a non-formal group. Therefore, from this point onward "organization" will be used in the same sense as "agency" was in text and resources above.

Time Banking Models and Sub-Divisions According to Sub-Systems: Taxonomy

This chapter presents a comparative view of TB in order to facilitate a systems analysis to be conducted in the following chapters. Based on a literature review using the main resources mentioned above as well as on my own research methodology described in the chapter "methodology," the main sub-versions of TB that have been identified will be described below, with the main divisions referred to as "pure models." These are the three models described above as "branches" (P2P, P2A and A2A), to which one more addition will be added:

- **Person to Person (P2P):** As described in previous section
- **Person to Organization (P2O):** As described in previous section
- **Organization to Organization (O2O):** As described in previous section
- **Organization to Organization Plus People (O2OP):** A new model of TB emerges out of a necessity to describe a situation in which all physical and legal bodies can participate and use a TB. Unlike Person to Organization, which consists of one organization and people, more organizations participate along with more people.

In the following sub-divisions (the white column in the table below), these main pure models will form a new TB taxonomy, which is based on my own research and provides the cornerstone for the further parts of this book. The sub-divisions reflect the systems analysis which follows, with the sub-groups also representing sub-systems of each of the aforementioned models of TB. The sub-divisions illustrated below are listed according to coordination scheme, service range, membership, base institution and time credit scheme. In blue are all sub-variants which would fit into all the pure models. The green fields represent special cases which can be adopted only under specific pure model approaches. The specifics of each of the sub-divisions are explained in Table 3.

Table 3. Possible combinations of pure models of TB with sub-systems

Time Bank Taxonomy: Possible Combinations of Pure Models With Sub-Systems					
Sub-Division Sub-System	**TB Model**				
	Pure Models	**Person to Person**	**Person to Organization**	**Organization to Organization**	**Organization to Organization Plus People**
Coordination scheme	Coordinator based	x	x	x	x
	Self-managed	x			
	Virtual	x			
Service range	Heterogeneous	x	x	x	x
	Homogenous		x		
Membership	With specific interest group(s)	x		x	x
	With focus on specific target group(s)	x	x	x	x
	Network of TBs	x	x	x	x
Base institution	NGO/Charity	x	x	x	x
	Governmental organization	x	x	x	x
	Business	x (possible, but uncommon)	x	x	x
	Non-formal group	x			
TC scheme	TB as a pension scheme		x		
	Monetarized TC (paper money, checks etc.)	x	x	x	x
	TC registered on a IT platform	x	x	x	x

Source: Author's research.

Coordination Scheme

Coordinator Based

Most authors agree that a coordinator (Boyle and Bird, 2014, Caroll and Belloti, 2015, Granger, 2013, Timebanking UK, 2005, Timebank Mahoning Watershed, 2017, Simon, 2004, Ryan-Collins et al., 2008, Naughton-Doe, 2015), sometimes called a broker, is essential for a TB to thrive (Marks, 2012). This does not mean that a TB can never exist without a coordinator, however most TBs in existence do in fact utilize a coordinator to facilitate exchanges. This is considered to be the most successful and safest model. A coordinator not only keeps the TB alive by actively connecting members (so called matchmaking), looking for new opportunities and services, organizing events (Valor and Papaoikonomou, 2016), but is also responsible for the safety of exchanges, networking, fundraising, keeping a database of members and their skills, knowledge, assets, resources and time (SKART – an extension of the previously introduced SKA; these elemental inputs will be summarized in next chapter), etc. In short, a coordinator is generally considered the essential element in TB, one which allows its smooth operation, safety and sustainability.

- **Advantages:** The coordinator keeps the TB alive, working and developing; a higher security level is maintained.
- **Disadvantages:** Higher costs, as coordinator is generally a paid position (a voluntary assignment may serve only as a temporary solution).

Self-Managed

A TB can also work without a designated coordinator. This is a situation where exchanges either do not need facilitation, the TB is organized enough to be able to meet and coordinate without a dedicated person, or the services are homogenous in nature (as described below). This model usually starts "bottom up" from a specific need in a certain environment, thus people draw together around a certain problem and would like to solve it by organizing themselves into a TB. Even a self-managed TB can have a "leader," i.e. one who informally takes responsibility for tasks and exchanges.

- **Advantages:** No need for a coordinator, although an interest group might have a "leader" who would motivate members. Low costs, and a bottom up approach means that the TB should remain self-sufficient until necessary; several self-managed TBs can join together to enlarge exchange possibilities.
- **Disadvantages:** Lack of coordination; one vision defines the existence of a TB instead of a set mission. Without a coordinator there is lower security as well as limited feedback about quality of services with self-management. Tends to often collapse.

Virtual

This TB can also be purely virtual, offering most of its services online. This is one of the special forms of a TB (Nikitina, 2011, Valor and Papaoikonomou, 2016). In this situation the role of a coordinator may be simply replaced by software or a web application. This IT solution provides a virtual "marketplace" where offers and requests are presented and, in addition, would also include the potential for feedback in the evaluation of exchanges. The lack of coordinator brings a lower level of security and control over the work of a TB.

- **Advantages:** Low running costs, no need for a coordinator.
- **Disadvantages:** Low security, usually depends only on online services. Collapses often due to a lack of security and lack of motivation, as there is no one actively connecting members.

Service Range

The majority of TBs may be considered heterogeneous, i.e. they have more than one strictly defined target group and various services to offer. Heterogeneity also indicates that the TB is open to diverse people and organizations which each bring their own SKART for exchange.

- **Advantages:** An abundance of SKART to exchange and share.
- **Disadvantages:** Higher demands for coordination.

A homogenous TB has members which provide only the same service or very similar ones. This is only possible in the P2O approach, by joining a network or as a result of being founded as a part of a network. This approach is usually applied when an organization requires a service to be done, which is then proposed, with members offering that particular service, forming a TC, which can then be used to claim rewards proposed by the organization (Marks, 2012). The situation can also come in reverse, i.e. when a TB is founded to activate people who then offer a service to an organization, through which a TC can also be used to great benefit. For example as HM Prison Gloucester (organization) does for its inmates (TB members) who refurbish bicycles, earn TC which can be given to their families to give them access to various services (Gregory, 2012).

- **Advantages:** Easy to be managed, direct work with particular target group, clarity of offer and demand. Actual problem solving.
- **Disadvantages:** The TB cannot exist on its own and must be integrated with another organization (P2O scenario, or in network).

Membership by Means of Specific Interest Groups

Interest groups of members often appear and grow within a TB, with members usually gathering around a common goal, idea, project or vision. The members engage in a particular endeavor, with the results usually coming through spin-off projects, innovations or even new TBs or branches thereof. The growth of interest groups within a TB is usually a sign of a healthy TB, and interest groups can be considered as multipliers of the benefits of a TB.

- **Advantages:** Multiplication effect of the benefits of TB.
- **Disadvantages:** Can cause disturbances in governance and in the division of power within a TB.

With a focus on one specific target group (or set of groups)

Certain TBs are dedicated to a specific target group or groups, with the latter actually the most common situation. A target group focus provides enhanced possibilities to a TB, as SKART can be predicted and the coordinator can work more efficiently with the group's needs and steer a TB accordingly. Target groups would also naturally be present in self-managed TBs. As groups develop from the bottom up, they would grow around and from respective target groups which may organically join through their mutual interests in certain offers and requests.

- **Advantages:** The wider capacity of SKART multiplies the possibilities of exchange and outputs and positive feedbacks of a TB.

- **Disadvantages:** The needs of a target group or groups should be respected

Networks of TBs

TBs can also join in networks. This can happen either by one TB spreading its branches, or by the conjoining of smaller TBs in a network in order to extend the range of SKART and with it the possible offers and requests. In a network, various types of interest groups in various shapes and sizes can emerge. Developments along these lines lead to extensions toward a higher level of Time Banking.

- **Advantages:** Massive growth of SKART and exchange possibilities.
- **Disadvantages:** Complex governance; the aims and objectives of networked TBs can differ.

Base Institution

Although a particular TB may consist only of a non-formal group of citizens, TBs in general tend to be legal bodies formed for various reasons. The most important factor is funding and institutional backing. It is believed that institutionalization is one of ways towards long-term sustainability of a TB (Papaoikonomou and Valor, 2017). Therefore, rarely can we find a group of people who have merely declared themselves a TB (although this is possible), rather we find a TB based on or around an institution of some kind.

Non-Profit Non-Governmental Organization (NGO)

An NGO is very common base institution for TBs, especially those of which that are founded by NGO representatives or by related independent activists. As these are non-profit organizations, this form is the most logical choice, allowing for a TB to become a sanctioned legal body. Depending on the laws of a particular country, NGOs usually engage in various activities which are not aimed at turning a profit, however generally they are also able generate their own funding in some ways. In addition, they can apply for various kinds of subsidies, beginning with governmental grants, through funding from different private foundations, to sponsorship from businesses. As a TB is always a not-for-profit organization, this legal form suits NGOs best and it is actually more convenient for governmental organizations or businesses to set up a spin-off NGO should they wish to engage in TB activities.

Governmental

Whether we are speaking about a municipality and its bodies or other governmental institutions, TBs can play important roles if legal system allows it, and if authorities are open-minded enough to see the benefits in time banking. The most common use of a TB in this case would be within the framework of P2O in which a government, generally a regional or local one, would like to empower its citizens through co-production in engaging certain activities. Also other TB models are possible when a governmental organization recognizes the benefits of using the TB concept.

Business

Although it is not very common, a business can be also part of a TB or a founding/base organization. The lack of popularity of TB among businesses is probably due to image of the TB as merely a social system. A typical business which would engage in TB would be a small to medium enterprise which is locally based. These may consist of, for example, restaurants or other small local businesses (Marks, 2012) conjoined in a TB in order to attract more customers, especially in times when their facilities lack customers. At times like these they may offer their services for TC partly as a marketing campaign, but also earning TC for their some of their operational costs. One good example of this use as well as of the monetarization of TC is the NGO Spice Innovations, originally located in Wales, an enterprise which also involves businesses in a TB scheme (Spice, 2017). Nowadays, with the vast development of social enterprises, TBs have even more to offer. Even considering more traditional businesses, the universal concept of TB can be used in groups of cooperating companies such as company clusters. As they cooperate, these units can keep track of shared resources in a TB kind of system where money is not required and the clarity of trade is recorded and tracked (Valek et al., 2014).

TIME CREDIT SCHEME

TC Registered on an IT Platform

The most common approach nowadays. A software or an application helps a TB coordinator or members (in coordinator-less approaches) to keep track of exchanges, quality (by feedbacking) and sharing information; the platform works as a record medium for TC, etc. The exact number of IT solutions is not known, as there is no official database, and any TB can create its own solution. Sources claim that there are about 30 of these systems in operation in the world (Boyle and Bird, 2014).

TB as a Pension Scheme

Examples exist in which TB is used as a pension scheme (Kennedy et al., 2012). In this set up, people perform work in order to obtain TC which are then stored for later use their pension, since time is not influenced by inflation. This means that in addition to a standard pension plan within a country a person can collect TC and use them to pay for services in a TB later in retirement. This may be particularly useful in environments where the financial system is unstable and a high rate of other financial instabilities exist such as inflation. Such a TB can also work as ordinary TB exchanging service with the prevalent storage tendencies.

Monetarized TB

Does an approach to TB exist in which TCs are issued as actual money? This in essence means that notes representing TC are physically printed and used as a currency, or an IT TB platform is opened to use for various businesses and other organizations, as written above. This approach is beneficial if the primary values of TB are kept, thus exchange is equal, as it broadens possibilities for members to "spend" their TC and opens new markets for businesses, also giving them something in return. But if the printed TC comes to be used in the same way as is money, in long run this would defy one of these core values, that of equality. Monetarized time credits would simply become monetary all over again, a situation that would mean that the added value of TB would diminish. We would once again be approaching the market economy, for which TB should be a complement, not a replacement. In lieu of this potentiality, it would be also possible that TB would turn to other CC approaches such as LETS or local currency, which, while proving beneficial for a community or local market, would lose the advantage of TB being purely a systemic tool supporting progress and innovation. In other words, TB aims to aid economies and societies on a systemic level by dealing with the origins of problems. Other approaches generally deal with the aftermath of consequences resulting from structural problems inherit in systems of government, economy and society.

The list introduced above stems from the research of the author to the date, but the solutions can never be final, as TB is a universal (Gregory, 2012), basic concept, one which can be adapted countless times, with future approaches remaining undiscovered. One of main objectives of this research is to demonstrate the universality of the TB approach, a strategy which requires the organization of classifications of TBs in a systemic way. The creation of this taxonomy will both facilitate the systems analysis to follow as well as determine common features among TB models, ultimately synthetizing the results into one universal approach.

Problem Situation Expressed

In order to identify the elements of TB, its properties and relations, a rich pictures approach will be used, as suggested in (Checkland, 1999). Rich pictures offer a way of capturing the complexity of human affairs, as this is always an issue with multiple interacting relationships (Checkland, 1999). This approach to the picture is defined by a particular point of view and offers a way to describe a problematic situation holistically. There are various guides to start with and in order to stay on track, i.e. to determine the focus of the rich pictures and the look for relationships between an internal situation and the environment (Tiger Spreadsheet Solutions, 2016). From the point of view of systems engineering, the main focus should be on the so-called 7 Ms (Skyttner, 2005): Messages, Methods, Measurements, Money, Machinery, Materials and Men. To concretize this more towards a soft system, in developing a picture of TB we can focus on:

- People
- Processes
- Environment (Checkland, 1993)

But it is also necessary to take into account resources, such as physical resources, skills, finances and time. As already noted above, SKA would be enriched by this approach by two more elements; resources and time. Thus, any entity which becomes a member of a TB usually commits the following to exchange: skills, knowledge, assets, resources and time. These can all be related to people, since TB as a soft system is composed of people, whether individuals or organizations. It might seem that funds have been omitted in this list, but as TB is non-monetary, investing funds is non-obligatory and usually funds are inputted from the side of the founding and/or base organization, or from another stakeholder.

Based on information above, the universal input into a TB system is SKART – Skills, Knowledge, Assets, Resources and Time. This list comes as one of primary results of the research and systems analysis, and these five elements are present in any exchange done in TB.

- **Skills:** Any entity, human or organization in a TB offers its skills directly or through people. Skills are the main carrier of the remaining elements of the SKART.
- **Knowledge:** It is required to have knowledge to manipulate any type of skill. Moreover, this knowledge itself can be offered or received in the TB scheme instead of or as a part of a service.
- **Assets:** One of core values of TB is that everyone is an asset, thus as a part of any transaction in TB people are assets. An asset in this understanding is any entity used within an exchange process, even – and mostly – people themselves.
- **Resources:** Although this might seem similar to assets, resources are more like "fuel" for the assets. Assets would be specific aspects including people, and resources are the means by which assets work properly.
- **Time:** Time is the primary variable that an entity dedicates to a TB along with all the aforementioned features. All organizations are composed of people and each person has in each day 24 hours to use for various activities. How much he/she devotes to a TB is up to him/her.

To give an example about SKART in a real TB situation, we might consider the service of driving someone to a destination in his/her own car. The skill in this sense would be the ability to drive a car; knowledge is that of the location of the destination along with the associated journey information. The asset is the car and a resource would be fuel, which leaves us with time we need to dedicate to the whole service.

Any physical or legal body has free SKART, which are not in use in a particular situation and are free to be exploited, although conditions might not allow it. If there is no demand or possibility to "sell" SKART, it would stay unused. TB facilitates the usage of these unused SKART, thus bringing benefits, which are non-monetary, but have a way of improving the life of an individual, society or organization(s). In a TB environment, SKART transforms into Offers and Requests. SKART enters as an input and in the transformation zone of a TB system, it changes to offers and requests, i.e. members use their SKART to define what they are able to offer, and what they would require (also based on what is available) in return, and the coordinator can match these factors accordingly.

Another conspicuous fact is that both the offer and request can lead to both the earning and spending of TC. One side offers, while the other simply matches to this offer and uses a service provided, so a member who offers something earns and the other party spends TC. Or, vice versa, i.e. one side requests something, the other side matches the offer, then the member who requested spends TC and the one who has matched to the request earns TC.

The Problematic Situation

The problematic situation was analyzed using rich pictures. The main problem is the lack of research which would support the wider use of TB concept across spheres of human activity. So far, the concept has been used mostly between people, and mostly in the so-called third sector, i.e. non-profit, non-governmental organizations. Businesses are rarely involved, and rhetoric used to promote the idea of TB gives it the appearance of having a solely social purpose. Starting from this idea approaches the concept of social enterprise, which is an ordinary business with added value to society (Social Enterprise UK, 2017). This concept may be involved in the business idea at all levels of its operation, whether products / services, employment, or use of earnings is considered. There is no reason which should prevent the utilization of the benefits TB offers in any sphere of human activity. A deeper analysis of the issue has shown that much of unused SKART goes to waste, since it is not marketable at the moment. Similarly, SKART of ordinary people goes to waste due to structural unemployment. TB provides a tool for people and organizations to help themselves and to use the SKART and by this, also bring benefits to society as a whole. The issues discerned by rich picture analysis have been mapped and divided into problematic ones and beneficial ones as shown below.

Problematic issues

- No relevant research (taxonomy, universal approach) to allow easy access and implementation of TB concept.
- TB is a systemic approach in its essence, but the neither the parts nor the whole have been unexplored and undescribed.
- Funds are needed to sustain TB; the cost is not prohibitive, but to run a sustainable TB it is necessary to have stable funding for at least the coordination stage.
- The issue of law and taxes usually depends on the understanding of benefits. Where understood that inputs are multiplied by co-production paradigm, the TB is exempted from taxation; where this is not understood the idea is usually killed and all benefits lost with it.

This leads to

- Unused non-marketable or marketable skills, knowledge and assets, or SKART, by all people, businesses and NGOs.
- Fear of something "illegal" being perpetuated due to the non-traditional approach, but when understood also from a legal point of view it brings benefits much higher than any potential "tax avoidance" which may be feared.
- Distrust from a business point of view, as profit is generally viewed exclusively as monetary. Other "softer" benefits such as in-kind, market share, promotion, image, sharing resources, and by these benefits lowering costs, etc. might not be seen as strong an argument as pure financial profit.
- Suspicion of tax avoidance might hinder the development of TB.

Benefits:

- In contrast to the market economy, by offering equal valuation it offers another innovative way of using SKART which would have been left dormant.
- It maps and records what might be a “grey” economy bringing non-monetary benefits by stimulating activities which would not happen without TB, ergo tax avoidance does not exist when there is nothing to tax in the first place.
- As a by-product of exchanges, various organizational and societal problems are solved.
- It activates unused markets, offering businesses new ways to operate.

Before a general universal system of a TB can be synthetized, the available approaches must be analyzed and broken down. Nine (9) approaches have been identified based on the taxonomy introduced above. Four of these main approaches can be called “pure models”: P2P, P2O, O2O and O2OP, which are then sub-divided into sub-variants according to particular model design and purpose (which also influences the presence and structure of various sub-systems). When compared, it becomes clear that it is not necessary to draw a model for each of sub-divisions, because a number of specifics may be introduced into a broader pure model. In Table 4, which is shown in the chapter featuring the TB taxonomy, TBs with similar content are marked in blue, indicating that all blue marked sub-systems may be included in all of the four pure models, and the ones marked in green can fit only into one. It was necessary to create all of the models, including the very specific ones. To summarize, there are four main “pure models,” which include main sub-systems, and also five specific cases. These would comprise self-managed and virtual TBs in a coordination sub-system: one of which in a membership range subsystem as a homogenous service TB; another of a TC scheme sub-system of the pension scheme use of a TB. And finally, a one membership sub-system, an idea for a TB network where the members of a TB are other TBs, with the scale in this model being the widest, as it can cover entire countries.

A Deeper Analysis of Time Bank Models and Sub-Divisions

As introduced above, all nine models from the taxonomy will be analyzed to maintain a systemic approach, although as indicated some are quite rare cases. In the proceeding pages diagrams and lists of elements are presented of each of the following approaches:

- Person to Person
- Person to Person: Self-managed
- Person to Person: Virtual
- Person to Organization
- Person to Organization: Homogenous
- Person to Organization: Pension scheme
- Organization to Organization
- Organization to Organization plus People
- Concept of TB networks

The pure models of TBs are the 4 above without text behind the colon (besides networks, which entails quite an abstract concept). As written in our Taxonomy, these models are the final result of merging the features of all known approaches to TB, with the aim of the analysis being to isolate all of their elements as systems and derive from these common elements of each. Based on these elements, we can then define the main core features common for all approaches, and design one universal model which would contain most of elements. Only those elements which would be considered to be totally irrelevant for the universal model is omitted, because these have only one purpose in a specific situation which can occur only in certain conditions in a certain model. Some elements must be kept within the core even though they are not present in all the models in exactly the same way. One example might be the coordinator, which would be “filtered out” from the core elements by a Virtual TB model, because it uses an IT solution to replace the coordinator, and in the self-managed approach, where this might or might not take place. Nevertheless, the coordinator holds such a strong and important role in all other models that it cannot be left out completely.

Diagram Design

The diagrams which are introduced below were developed on the basis of a circular design of P, Q, R as introduced above. Their origins are in multiple iterations of application of SSM in the environment of the TB taxonomy. Some selected diagrams from the drafting period are included in appendix and they illustrate the complexity and development of the final simplified diagrams and also the main elements listed in the charts below. The original diagrams were done with paper and pencil, as rules for the rich pictures suggest, but for a clearer presentation they were re-done with the help of Microsoft Office. They were simplified to illustrate the main relations and elements while still keeping a level of clarity, and they represent a schematic view of a particular TB model, which is further supplemented by charts and explained in greater depth.

Following the proposed circular model, there is always a circle of system P where transformation happens. The transformation includes sub-systems Q and their sub-systems Q2, which involves both actors and customers. Around this there is the wider system R where the entities on the ownership level can be found, as well as environmental influences. Each model differs from the previous one, but they maintain the main sub-systems of base organization (or equivalent) and coordination, membership, time credits which are usually drawn in P keeping the same rules and placement for higher clarity. In the original models of TB three colors of arrows are present: blue ones represent ordinary processes, whereas green stands for feedbacks, and orange are inputs. Inputs and outputs are not strongly underlined, as they are described in more detail in the charts and in the breakdown. The colors will differ in next chapter, in which the universal model is introduced with emphasis on processes, information flows and exchange of knowledge. These are marked with different colors.

Chart Design

The charts are divided into several sections. Because of the difference in TB approaches, these sections vary in size and shape. This is natural, as we are dealing with a soft system and, as much as we would like to maintain a unified view, this is not always possible. The chart contains systemic elements isolated from the rich picture design which preceded this phase, and which is based on the circular design and approach of P, Q, R.

Starting from the left, the column with Inputs is featured. Inputs are entities which come and are provided by the outside environment, i.e. from the wider system R into the system area P, which is the place where transformation takes place. P is divided according to its sub-systems Q, which are always divided in the sub-system entities (in some cases also their sub-systems Q2), and these are followed below this by their own processes. On each side, specific columns have been designated FB which indicate at certain places if there is positive or negative feedback from particular process outside of P to R.

At the bottom of the chart there is the domain of the wider system R, where the entities and processes they undergo in relation to P and its sub-systems Q are also listed. Finally, in the column on the far right can be found the outputs of the system. Outputs are inputs that have been transformed in P and are divided into three parts: Impacts on R entities, i.e. outputs which impact directly on entities in the wider system; Systemic Outputs, i.e. outputs which impact the outside environment and which may or may not also include R entities, but have wider systemic impact far outside of the system P; finally, Systemic Feedbacks strengthen or weaken specific element(s) within or outside of system.

In addition, certain specific terms and abbreviations should be explained. ETQOR stands for Exchange_Time Credits_Quality_Offers and Requests. The often-used abbreviation TG stands for "target group." The commonly-used term "citizens," which is also the name of one of the entities, may be understood as people, or human beings, rather than in relation to national citizenship. The expression "societal" is also often used in this context, understood as related to the whole society, i.e. people, organizations like businesses, NGOs, government(s), the outer environment of human activities, etc. IT Solution relates to any computer-based system (software, web application etc.) which serves as a record medium for exchanges and as a base for an exchange platform. This means that it also might host other features if necessary and if available, like the database of SKART, matchmaking elements, etc.

Each chart has at its lower base slightly detached cells with specific information. These cells contain information about the TB Network model which is very general. It is not significant for this chapter, but it ought to be noted to maintain a complete overview of TB as a system.

Systemic Decomposition of TB Elements: Inputs

To keep the following chapter concise and clear, now all the mentioned inputs, entities and their processes compiled from all TB models will be explained. To fully understand the following charts knowledge about each element is necessary. After this, each of the nine models will be introduced along with its own chart and the main differences among them will be emphasized. Outputs will be summarized after the analysis, as they are more clearly represented in the charts.

Box 1 illustrates the structure of all the inputs which were isolated in the analysis, with each being explained. In blue are marked the ones which are rare cases.

Funding

One of most important inputs from the wider system (R) to any TB system is funding. In most of cases, any of the R entities can fund a TB. The term funding might sound misleading considering the way TBs operate, but the idea here is related only to funding a coordinator role and an office (which may be provided in-kind by a base organization), or meeting point and possibly equipment (in this case this can be also provided in-kind and would be part of physical resources). Funding also might include webhosting, web design, web administration, etc. In rare cases it would include software development, but this would usually be not necessary, as many options of free IT solutions exist. It also might include the fee for being part of a TB network. Anyway, in

Box 1. Full list of inputs from all nine models

<table>
<tr><th colspan="3">Full List of All Inputs From All Nine Models of TB</th></tr>
<tr><th>Input</th><th colspan="2">Variant of Input</th></tr>
<tr><th colspan="3">Funding</th></tr>
<tr><td>Founding</td><td>Founding with a specific goal</td><td>Funding only by NGO or Non-formal Group of citizens</td></tr>
<tr><th colspan="3">Joining</th></tr>
<tr><td>Recruitment (people)</td><td>Recruitment (organizations)</td><td>Recruitment (organizations and people)</td></tr>
<tr><td>Unused SKART</td><td colspan="2" rowspan="8"></td></tr>
<tr><td>SKART</td></tr>
<tr><td>Law</td></tr>
<tr><td>Physical resources</td></tr>
<tr><td>Rewards offer</td></tr>
<tr><td>Activities offer</td></tr>
<tr><td>Activity offer</td></tr>
<tr><td>IT Solution admin</td></tr>
<tr><td></td><td>Whole different TBs Join with their members, target groups, interest groups, SKART etc. (All kinds of previous models)</td><td></td></tr>
</table>

Source: Author's research.

most cases the funding involves only one or two salaries and minimal office costs. To summarize, the funds inputted into TB system are necessary for its sustainability, but this is a very small sum, especially compared with the value of outputs.

Founding

This R entity founds a TB (in reminder, R is ownership level of the SSM). Whether it would be a NGO, governmental organization, business, or non-formal group of citizens, possibly a main input would be to actually a start the system itself by providing it with rules and potential statutes. Which organization this would be depends solely on the situation, the aims of a particular entity, and problem that needs to be solved. This input has also its variants. There are two cases of P2O models in which the TB is founded with a specific goal; or, we can say, a mission. This goal defines the range of services which may be transformed in TB system, and the founding organization is usually a governmental organization, sometimes a NGO. A second variant is a situation where TB would be founded only by a non-formal group of people, which can also form a NGO. This is the situation of the P2P Self-managed TB model.

Joining

Joining is also one of main inputs. This describes the situation in which a person or an organization becomes a member, thus a part of a membership sub-system. As will be presented in each separate case, in each model there may be a number of barriers for some entities to join, or a specific model is reserved only for a specific member group. In a broader sense, a TB may be aimed at only one target group, but in this analysis we would only take into consideration whether a model is open to being joined by only one target group or by multiple groups. This is shown by listing it among membership entities.

Recruitment

This differs from joining by the fact that it is directly influenced by the feedback from the system or, more precisely, mainly by the activity of the coordinator and security matters. It can be also influenced by the feedback of members regarding the amount and quality of services, ETQOR, or simply pure information about the activity and range of Offers and Requests. Anyway, although direct feedback by members is important, this feedback will generally be facilitated again by the coordinator. Recruitment has a couple of other variants, as there may be limitations, such as above in the section Joining. In general, it would consist of divisions of people, organizations or both.

Unused SKART

As introduced above, SKART is one of main inputs which "fuels" a TB. People offer their free/unused SKART, which are transformed in a TB system into offers and requests, and in this way exchanges happen. Unused SKART are non-marketable Skills, Knowledge, Assets, Resources and Time which would go to waste. By inputting it into a TB system, valuable outputs are produced.

SKART

In rare cases, members can also offer their SKART, which are then put into use; in other words, this would be work they sell on the labor market. This would solely depend on how much value the service(s) available in TB has for them, or the possible rewards in P2O models. Still, this possible input should be also taken into account.

Law

Governmental laws can allow, support, or restrict or forbid a TB. Thus, in all models the role of government is marked by both positive and negative feedback. If a government understands TB in its systemic role in the economy and society and provides a legal framework, it is usually a win-win situation and even governmental organizations often run their own TBs. But the opposite can also happen. By seeing TB through the lens of a market economy, and as an alternative rather than a productive complement, it may be misunderstood. Laws can be enacted that disrupt and make impossible the founding or running of a TB, as issue introduced in the earlier chapter "Controversies regarding law and taxes."

Physical Resources

Aside from funds, physical resources may be provided from the outside environment of the wider system. This may take the form of various in-kind contributions such as office space, office materials, transportation and practically anything which would be of use for a TB or its development. This may also provide the space for organizing events, resources for spin-off projects and supporting innovations that come as an output. To this category can be also added IT solutions, which are one of cornerstones of working in TB in these times. These generally come at no financial cost, and represent one of the key resources.

Rewards, Activities and Activity Offers

These are the sole domain of P2O models. They are founded with a goal and they also provide work offers, here called Activities, (one activity in the case of the homogenous model) and offer rewards. Activities allow the earning of TC, and Rewards are the possibility of redeeming them.

IT Solution Admin

This is an input unique for the Virtual P2P approach. An administrator is hired or enlisted to keep the system running, or at least to implement it and then leave the maintenance to the members.

SYSTEMIC DECOMPOSITION OF TB ELEMENTS: ENTITIES AND PROCESSES OF THE SYSTEM

The main elements of system P which transform inputs to outputs are:

- Membership
- Base organization (or equivalent)
- Coordination
- Time Credits

The division of these sub-systems is not random. It is result of many iterations of rich picture design of various approaches to a TB. All approaches in the Taxonomy have sub-systems which would fit into this category. This division is therefore based on reality and is a basis for analysis so universal elements can be identified. Only on the basis of these elements can one universal approach be created which would join most of the features, but may in the end take quite different forms.

Entities

One TG and Multiple TG stand for one or multiple Target Groups within a TB, i.e. whether a TB is focused on (their members come from) only one target group or multiple. If both entities are present, it means that TB may have a members' base made of only one target group, but it can also grow into a multiple TG. If there is only the multiple target group entity, it would mean that the TB is open to anyone. A target group may be comprised of for

Box 2. Membership subsystem

Membership	
Entities	**Processes**
One TG	Focus on TG(s)
Multiple TG	SKART info
Interest groups	Multiple TG SKART INFO
Offers/Requests	Feedback about (ETQOR)
Leader(s) (non-formal coordinator)	ETQOR Feedback only by exchange platform
Givers	Info about O+R
Takers	FB from members about rewards usability
Members are organizations from R: NGO Business Governmental org. Citizens	Select activity
Various TBs connected to exchange platform	Select the one activity
	Giver change to Taker over long time
	Form interest groups
	Exchanging information about O+R
	Exchanging information about SKART

Source: Author's research.

example students, a retirement community, entrepreneurs, restaurants, social enterprises, SMEs in a region, etc.

Interest groups are important part of TB development. These might be formed by the coordinator, but most of the time they form around a particular situation or a problem to solve, with the results often being innovations and spin-off projects. This grouping of people around an issue is very similar to what we know from knowledge management as communities of practice, where people informally gather in the business sphere around an issue to be resolved (Manville and Foote, 1996). The same situation happens in TB and interest groups can become most important in terms of output in innovations.

Offers/Requests, as indicated earlier, are formed from inputted SKART. Members are carriers of SKART, which is transformed by input into the system into Offers and Requests, and therefore can be considered an entity of Membership.

Leaders are non-formal coordinators in the self-managed P2P model (the entity may be present also in P2P virtual, but it is not common, as members might never meet each other in person). He or she is simply a regular member who has assumed a leadership role. This brings some of the benefits of having a coordinator, but as this is not a formally appointed person based on rules or statutes, the individual does not carry as much weight in terms of decision-making power.

Givers and Takers are entities in the pension scheme P2O model. Members who have been giving (providing services) and saving TC over time become takers who are withdrawing their TC "savings." This typically happens over long period of time (typically a transition which comes at pension age).

Members are regular bodies, physical or legal, who enter from the wider system R into the system of a TB and are transformed to members. Members can belong to one or more target groups; they form interest groups and they have their offers and requests based on their SKART.

Various TB is a conceptual entity of a model of networks of TBs where members are actually complete TBs of various kinds.

Processes

Focus on a TG - see paragraph on target groups above.

Form interest group is a process in which a member or group of members starts an interest group to resolve some particular issue or to engage in a new activity. The interest group can be also open to the public, i.e. for people or organizations who are not yet members.

SKART info and Multiple SKART info are processes by which members provide information about their SKART to a coordinator so he/she knows what are capabilities of members in order to provide successful matchmaking. The difference in singular and multiple SKART is related to the focus on target groups. In this sense, pure SKART info refers to a situation of information coming from one target group, whereas the Multiple variant comes from several target groups. This division underscores the fact that each new target group multiplies all five elements of SKART and, with this, also the range of offers and requests into which they can be transformed.

Feedback about (ETQOR) as explained in the above exchange, TC, quality and offers and requests are one of main flows going through a TB. Feedback flows outside of the system into the wider system and thus influences potential joiners. The attractivity of participation in a TB is influenced by the smoothness of exchanges and frequency thereof, the possibility to gain and use TC, quality standards, which include safety and a variety of offers, and also requests. This feedback, which can be both positive and negative, directly influences the enrollment of new members and recruitment. The coordinator can use this actively to share non-private information and statistics, thus providing transparency. In addition, members tend to provide this kind of feedback to their peers in wider system.

ETQOR Feedback only by exchange platform is an analogic process as described in the paragraph above, with difference being that this is associated with specific cases of P2P Self-managed and P2P Virtual models. Neither of these models has a coordinator and all the feedback flows only by exchange platform in a passive way4 much like the next process described. But unlike the following model, this one contains as much complex information as possible. Again, feedback can be both positive or negative, depending on how things are going in a TB.

Info about O+R might seem similar to the previous process, but unlike the complex feedback above, this is only information. It is static result of the transformation of SKART into offers and requests which are publicly available on an exchange platform of a TB. It can also provide positive or negative feedback, but unlike the previous active process this one is passive and relies totally on potential members to check the exchange platform. It is very important to notice that both the offer and request can lead to either earning and spending of TC. One side offers, and other matches this offer and begins to use a service provided, so a member who offers earns, and the other party spends TC. Or vice versa, one side requests, the other side matches, then the member who requested spends and the one who matched the request earns TC.

FB about rewards usability relates solely to P2O models in which rewards are provided by the base organization, or partner organizations, to provide the possibility of using TC. This feedback can again be positive or negative, depending if it is worth it for people to join a TB and become members.

Select activity and Select the one activity again signify a specific process when people join and become members in P2O models. Joiners have the option to choose an activity or a job to be done for TC. This activities are proposed by the base organization and follow the goal for which it was founded. In the homogenous model only one activity is proposed.

Giver change to Taker over long time is a long-term process which can be considered a delay between earning and using TC in the P2A pension scheme model.

Exchanging information about O+R and Exchanging information about SKART are general processes which happen between whole TBs in a TB network, the nature of which is understandable from the paragraphs above.

Box 3. Base organization or equivalent sub-system

Base Organization or Equivalent	
Entities	**Processes**
Statutes	Coordinates (provides coordinator)
Rules	Builds rules and statute upon a law
Database of Offers and Requests	Decision making
Matchmaking element in the IT Solution	SKART INFO receiving
Proposed Rewards	SKART INFO use for matchmaking
Coordination (is Q of Base.org)	Accessing the SKART database input
A decision making sub-system	Accessing the SKART database read
A decision making body	Matchmaking
	Activities proposal
	Activity proposal

Source: Author's research.

Entities

Statutes and Rules are one of most important aspects of the base organization, which is a legal body and must adhere to local laws and regulations. On this basis, the rules of a TB are constructed which provide security and a framework of operation for the members, a code of conduct along with any other specific rules to secure TB functions and sustainability.

Database of Offers and Requests is a database which is a part of an Exchange platform with the P2P Self-managed and Virtual models. Members directly input their offers and requests here.

Matchmaking element in the IT Solution. Without a coordinator there has to be inherited feature within a TB IT solution to provide the same function. This feature might be replaced by a simple "search" function which leaves any matchmaking completely in hands of members.

Proposed rewards are in case of P2O and P2O a homogenous kind of database or list where possible rewards are listed. Therefore, they are presented as entities rather than processes.

Coordination (is Q of Base organization). As Q means sub-system, this entity is actually a whole coordination sub-system under the base organization. This is situation which occurs in most cases.

Processes

Coordinates (provided by coordinator) means that a base organization provides, or finds, a coordinator for a coordination sub-system.

Builds statute upon a law relates to the first entity. This is the process of founding a TB according to a law, which is role of a founding entity.

Decision making is in hands of founding base organization. The coordinator is responsible for coordination, but as a legal body, the base organization is responsible for maintaining a body which governs it. In some cases, the legal representative and coordinator can be the same person, but this is not a practical solution.

SKART INFO receiving, SKART INFO use for matchmaking. SKART information is received from members and stored on the exchange platform in the case of the P2P Self-managed and virtual models, or in a database of SKART in other models. This information is later used for matchmaking. Due to the transformation process, at the moment the information is recorded in a platform or database, it may be also defined as an offer or request. The main difference is that SKART contain information other than offers or requests, as the abbreviation implies, but offers and requests are built on these skills, knowledge assets, resources and time; they reflect it and are derived from it. This broader information is used for matchmaking.

Matchmaking is an activity where a coordinator finds the appropriate members based on SKART or later on the offer and request (but that they can do also themselves). The importance of matchmaking is matching the right members pro-actively so they can start an exchange, but also to be able to facilitate the exchange of already-provided offers and requests based on the SKART provided as supplementary information. Based on offers and requests, members are also able to match among themselves, which is how it works in the P2P Self-managed and Virtual models.

Activities (Activity) proposal this is a one-time process of proposing an activity or activities in the P2P (homogenous) model.

A decision making sub-system and body are parts of a TB network which comprise a sub-system that contains some kind of decision making element, such as a board of members, a general assembly, etc.

Coordination

Coordination is usually a sub-system of the base organization. It is responsible for a variety of functions which are performed by coordinator(s) and other entities involved.

Box 4. Coordination sub-system

Coordination	
Entities	**Processes**
Coordinator	Matchmaking
Database of SKART	SKART INFO receiving
Leader(s) (non-formal coordinator)	SKART INFO use for matchmaking
Coordination	Accessing the SKART database input
	Accessing the SKART database read
	Events organizing
	Networking
	Fundraising
	Managing
	Security keeping

Source: Author's research.

Entities

The coordinator is the center point of a TB. He/She manages the SKART database, matches according to this offers and requests and/or members who have not participated yet. He/she also oversees exchanges, keeps security and conducts most of activities which are listed in processes in order to keep the TB secure, alive, prosperous and sustainable. More than one coordinator can be utilized, usually depending on the size of a TB. Considering the list of activities below, the work might be quite demanding for just one person.

Database of SKART may be digital, but also a kind of paper record, depending on the technological progress in the area where a TB is implemented. Data contained are based on information provided by members. This may vary but in general it reflects SKART, but also can contain other personal data like age, region, size of an organization, general area of operation, or interests in the case of people, etc. This database helps to find precise entities which might have free SKART to cooperate with, thus making the variety of offers and requests bigger. It is also used for security keeping, because it can provide information which entities can both exchange and cannot exchange based on various reasons.

Leader(s) has been explained above in the membership sub-system, because they are ordinary members. They are also noted here, as they might or might not perform the role of coordinator, so they belong to this category as well.

Processes

Matchmaking, SKART INFO receiving and SKART INFO use for matchmaking have been explained above, where it was described in relation to situation in cases when these processes are performed by an exchange platform or a non-formal leader. At this point they can be fully utilized by a coordinator.

Accessing the SKART database input and read is a process by which a coordinator meets with a member entity and inputs SKART information and additional data into the database. "Read" is a situation when a coordinator searches for information to match. SKART and other data can be also inputted into a database by members directly when an appropriate IT solution is available. Even though members can input, they cannot read, at least not most of the information. Some of information would be protected by privacy rules and only the coordinator can read everything to make an informed decision about matchmaking. This also limits self-managed, and especially, virtual TBs in terms of the amount of information they can go public with.

Events organizing, networking. One of jobs of the coordination entity, is to keep the membership base alive. For this, various events are often organized, mostly with the aim of networking, but also for other occasions. Networking can happen, of course, in other ways, but one of the ideal ways is to organize it, for example, a social event where member entities (or their representatives) can meet directly and exchange possible ideas about cooperation and exchanges.

Fundraising is mentioned in input section, although the costs of running a TB are really almost insignificant, funds are still needed to ensure sustainability, and it is usually the job of a coordinator to find such funding if it is not sufficiently provided by the base organization.

Managing relates to the general management and administration of a TB which is necessary to keep a TB together and running. It also relates to information management, in which the coordinator's role is to facilitate the flow of information through all entities (we can even call these the nodes of a system), as will be introduced in the discussion chapter.

Security keeping is basically an activity of the coordinator where he/she checks to make sure members adhere to the rules adopted by statute, therefore built on law. It also means that coordinator considers the capabilities of various member entities and facilitates contacts accordingly.

Time Credits

A Time Credits sub-system contains all the entities related to the main exchange medium; an equal Time Credit. A TC sub-system is like the blood circulation system of a TB.

Entities

Time credit supply is usually the virtual "register" of existing TC; this register can also be in paper form if information technologies are not available.

Exchange platform is usually a sub-system of TC except in the cases of the P2P Self-managed and Virtual models, where it replaces the base organization and coordination. An exchange platform contains a record medium, which can take the form of an IT Solution; it may be a kind of ledger book to keep records, or actual paper TC can be issued in the form of "kind of" money.

Box 5. Time Credits sub-system

Time Credits	
Entities	**Processes**
Time Credit supply	Earning TC
Exchange platform: Sub-sub-system	Spending TC
Database of SKART: Record medium: IT Solution Ledger Paper TC Offers/Requests Offers (givers) Requests (takers) Administrator	Earning TC (giver)
	Spending TC (taker)
	Work provided
	Work received
	Approval of TC
	Issue of TC
	Use of TC
	Giver save
	Taker withdraws
	FB about transactions to coordination
	FB about transactions to exchange platform
Record medium (a sub-system)	
	Processes: Use of TC Issue of TC FB about transactions to coordination

Source: Author's research.

The ramifications of monetarizing TC is explained further in section TB Taxonomy above.

The Record medium sub-system also contains offers and requests descriptions to provide information to members about actual possibilities to earn TC or to Spend/Use them. Offers and Requests are also divided into one sub-division in terms of givers and takers in the P2A pension scheme, where there is a difference between members who provide offers and who request. And finally, there is the Administrator entity, which relates to the P2P Virtual model where this individual can be also the IT administrator.

Analogically at the bottom of the chart is the outline of TC exchange in terms of Networks of TBs where a record is kept about the use and issue of TC, along with feedback to the main coordinating entity to provide an overview of all transactions.

Processes

Earning and Spending of TC including Giver/Taker scheme is quite a clear process just from the name. TC earned by finishing a task which was offered or requested in a TB brings a gain of TC which can be spent/used again for Offers and Requests. One important fact is that both earning and spending brings positive feedback by means of successful matchmaking, growth in social capital, and most importantly, because each transaction multiplies inputs in transformation through TB system to outputs. The Giver/Taker scheme presents the same situation, just taking into account the time delay between Earning and Spending.

Work provided and Work received is the information for a coordinator or exchange platform that work was done.

Approval of TC refers to the fact that in some special cases the TC record must be approved before it is released. The relates to P2O models, where the work has to be checked to make sure it has been done before TC can be issued.

Issue and Use of TC means that besides Earning and Spending TC, which consequently leads also to Issue and Use of TC, the TC can be also issued and used outside of exchange. This is common in P2O models, where members firstly earn, and later spend for rewards. But this can be also used as a practical tool. TC can be issued in limited amounts to stimulate exchange or they can be given "free of charge" to members to try new services and to make them active. This is usually used as a start-up package. Or TC may be used, for example in the case that a member leaves without spending/redeeming them, dies, goes bankrupt, etc. Then TC are left which can be used for various aims. TC can be also donated for later use.

Giver save and Taker withdraws is a process related again to the P2O pension model. This reflects the time delay between earning and using TC, i.e. when they are "saved" for a certain period of time and lie dormant and used later after withdrawal.

FB (feedback) about transaction to the coordinator or exchange platform represents feedback about transaction/exchange that has occurred to the coordination entity. This is usually done by "logging" TC from parties on both sides of the exchange, so one side logs the exchange, and second verifies that this amount of TC was earned or spent.

WIDER SYSTEM ENTITIES

There are five main wider system entities in all models. Firstly, citizens, e.g. ordinary people not engaged in any organization, can form into an organization or form non-formal group which can found a TB. There are then NGOs, or non-profit, non-governmental organizations which can found or fund a TB, provide resources for it or became a partner organization to provide rewards in the P2O models. Businesses are in same position as NGOs. Governments through governmental organizations can also perform all the previous activities, but aside from this they provide the law framework for TBs to form their rules upon. And finally, Involved TBs proceeds again from concept of the TB Network, in which complete TBs from the outside environment enter the system P.

Box 6. Wider system entities

List of All Wider System Entities		List of All Processes	Exceptions
Citizens		Join, resources, can fund by own funds, can found base organization	Are just affected by impact of TB in O2O model
NGO		Founds, Funds, resources, can become a partner org	
Businesses		Founds, Funds, resources, can become a partner org	In P2P Self-managed they can be involved but usually are not
Governments	Governmental org.	Founds, Funds, resources, provides law frame, can become a partner org	
	Law	Provides legal framework for rules and statutes of TBs	
Involved TBs		Joins	

Source: Author's research.

Summary: Citizens, or people, have most variability, and power. They can not only join existing TBs and found the self-managed model of TB, but also they can join or found an organization which would become a base organization. This finding maps to previous assumptions which are rooted in the theory of knowledge management, as organizations, whether NGOs, businesses or governmental bodies, are "powered" by people (Lesser and Storck, 2001). TB in this case becomes a tool for problem solving across sectors.

REFERENCES

Blanc, J. (2011). Classifying "CCs": Community, complementary and local currencies' types and generations. *International Journal of Community Currency Research*, *15*(D), 4–10.

Boyle, D., & Bird, S. (2014). *Give and Take: How timebanking is transforming healthcare*. Timebanking UK.

Bureš, V. (2011). *Systemove mysleni pro manazery*. Pribram, Czech Republic: Professional publishing.

Caroll, J. M., & Belloti, V. (2015). Creating Value Together: The Emerging Design Space of Peer-to-Peer Currency and Exchange. In Collaborating through Social Media - CSCW 2015. Vancouver, BC, Canada: ACM.

Checkland, P. (1993). *Systems Thinking, Systems Practice*. Chichester, UK: John Wiley & Sons Ltd.

Checkland, P. (1999). Soft systems methodology: A thirty year retrospective. *Systems Research and Behavioral Science*, *17*(S1), 11–S58. doi:10.1002/1099-1743(200011)17:1+<::AID-SRES374>3.0.CO;2-O

EEA. (2018). *Members*. Available: https://entethalliance.org/members/

Granger, P. (2013). *Valuing people and pooling resources to alleviate poverty through Time banking. Lewisham*. London: Rushey Green Time Bank.

Gregory, L. (2012). Time and punishment: A comparison of UK and US time bank use in criminal justice systems. *Journal of Comparative Social Welfare*, *28*(3), 195–208. doi:10.1080/17486831.2012.742452

Kennedy, M., Lietaer, B., & Rogers, J. (2012). *People money: The promise of regional currencies*. Axminster, UK: Triarchy Press.

Lesser, E. L., & Storck, J. (2001). Communities of practice and organizational performance. *IBM Systems Journal*, 40.

LMC. (2017). *Research proposal flowchart*. Available: http://irclmc.edu.np/research-proposal-flowchart/

Manville, B., & Foote, N. (1996). Harvest your workers´ knowledge. *Datamation*, 7.

Marks, M. (2012). Time banking service exchange systems: A review of the research and policy and practice implications in support of youth in transition. *Children and Youth Services Review*, *34*(7), 1230–1236. doi:10.1016/j.childyouth.2012.02.017

Molnar, S. (2011). Time is of the essence: The challenges and achievements of a Swedish Time Banking initiative. *International Journal of Community Currency Research, 15*, 13-22.

Naughton-Doe, R. (2015). *An evaluation of timebanking in England: What can timebanks contribute to the co-production of preventive social care?* (Dissertation). University of Bristol.

Nikitina, N. (2011). *Manual of a client of a time bank* [Настольная книга клиента банка времени]. Nizhny Novgorod, Time Bank Nizhny Novgorod [Банк Времени Нижний Новгород].

Papaoikonomou, E., & Valor, C. (2017). The Institutionalization of Alternative Economies: The Processes of Objectification in Time Banks. *Journal of Macromarketing*, *37*(2), 167–179. doi:10.1177/0276146716672286

Ryan-Collins, J., Stephens, L. & Coote, A. (2008). *The new wealth of time: How timebanking helps people build better public services*. London: New Economics Foundation.

Shreiber, G., Akkermans, H., Anjewierden, A., Hoog, R. D., Shadbolt, N., Velde, W. V. D., & Wielinga, B. (2002). Knowledge Engineering and Management. Cambridge, MA: The MIT Press.

Simon, M. (2004). *On becoming a time broker*. Gloucester, UK: Active Communities.

Skyttner, L. (2005). *General systems theory: problems, perspectives, practice*. Hackensack, NJ: World Scientific.

Social Enterprise UK. (2017). *Social Enterprise*. Available: http://www.socialenterprise.org.uk/advice-services/sector/social-enterprises

Spice. (2017). *Spice Innovations Ltd*. Available: http://www.justaddspice.org/

Tiger Spreadsheet Solutions. (2016). *Rich Picturing: Six Tips for Getting Started*. Available: http://tigerspreadsheetsolutions.co.uk/rich-picturing-six-tips-for-getting-started/

Timebank Mahoning Watershed. (2017). *Time Bank Models*. Available: http://tbmw.org/learn-more/time-bank-models/

Timebanking UK. (2005). A bridge to tomorrow: time banking for baby boomers. Gloucester, UK: Timebanks UK.

Timebanking UK. (2015). *Time Banking types*. Available: http://www.timebanking.org/what-is-timebanking/what-is-timebanking/

Timebanks USA. (2017). *About Timebanks USA*. Available: https://timebanks.org/timebanksusa/

Valek, L. (2016). Systems Archetypes in Time Banking: The First Step. *Vision 2020: Innovation Management, Development Sustainability, and Competitive Economic Growth*, 2052-2058.

Valek, L., Kolerova, K., & Otcenaskova, T. (2014). Time banks and clusters: Similarities of sharing framework. *Global Journal on Technology, 6*, 31-36.

Valor, C., & Papaoikonomou, E. (2016). Time Banking in Spain: Exploring their Structure, Management and Users' Profile. *Revista Internacional de Sociologia*, *74*(1), 14. doi:10.3989/ris.2016.74.1.028

Chapter 4
TB Models

ABSTRACT

Results of the systemic analysis conducted in the previous chapter are presented here. These are introduced as a TB taxonomy comprising nine types of TB. Each TB type is described with help of graphical models, specifically diagrams and charts. Diagrams provide graphical representation, and charts are presented in a tabular form in which particular specifics are described. The final part of this chapter discusses root definitions for every single TB model. In the end, common outputs are presented. Consequently, based on all available information, theoretical background, and the analysis of current approaches, a universal TB model is proposed. Similarly to single models in the TB taxonomy, the description of main characteristics such as inputs or subsystems of this universal model is provided.

CHARTS AND DIAGRAMS OF TIME BANK MODELS

This sub-chapter introduces charts and diagrams for each of the TB models. For each model there is a commentary about the main differences from other approaches and main differences in inputs, the system P, its sub-systems Q and Q2 and wider system R. Also, outputs are present this time on the charts. Outputs will be briefly commented on and a summary is presented at the end of this section. The models will be introduced in a logical progression from simplest to most complicated in P2P and its special models with Self-management and Virtual, through P2O and its sub-versions of Homogenous approach and Pension scheme, to O2O and its higher version O2OP. Lastly, the concept of TB Networks will be introduced.

DOI: 10.4018/978-1-5225-6974-9.ch004

Person to Person TB Model

The Person to Person TB Model is shown in Figure 1 and Box 1.

Overview

The P2P TB model is a basic one, usually founded by active citizens through a NGO.

- Inputs are visible in the left column. The only special feature here is that recruitment and joining is open only for people. Even if a person is an entrepreneur, thus a legal entity, it enters P2P TB as a human being (otherwise we would consider such person an Organization, and it would be entered in a different model).
- The System (P) and its Sub-systems (Q, Q2). The system contains all available sub-systems, Membership base, Base organization. Coordination is sub-system of the base organization, as well as Time Credits with the sub-system of Exchange platform.

Figure 1. Person to Person model
Source: Author's research.

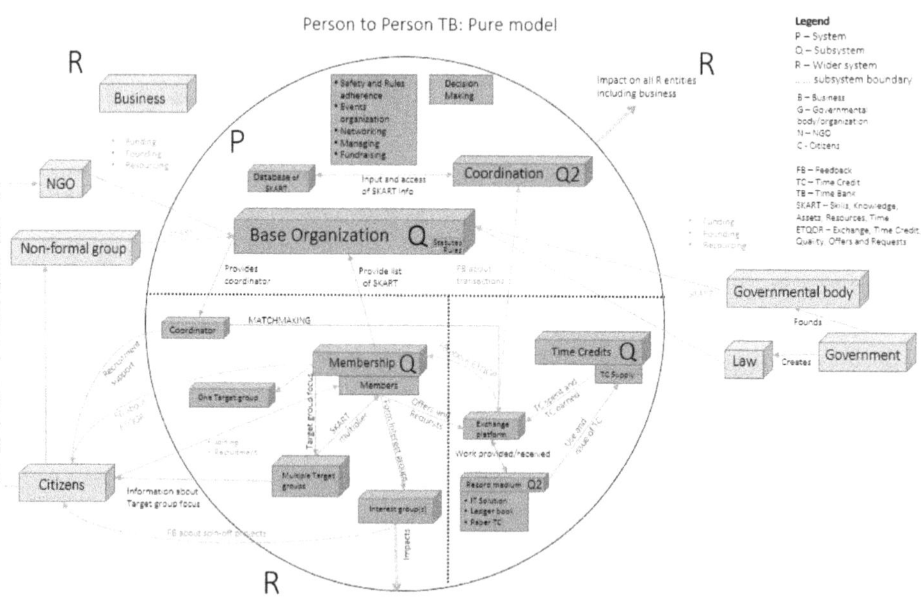

Box 1. Person to Person model of TB

Person to Person pure model of TB							
Inputs	**FB kind**	**P: The system (Transformation process)**					
Funding		**Q: Sub-Systems**					
Founding		**Membership**	**Base Organization**		**Time Credits**		
Joining (people)			**Coordination (is Q within base org)**				
Recruitment (people)		**Entities**	**Entities**	**Coordination Q2**	**Entities**		
Unused SKART		One TG	Statutes	**Entities**	Time Credit supply		
SKART		Multiple TG	Rules	Coordinator	Exchange platform	Sub-sub-system Q2	
Law		Interest groups		Database of SKART		Record medium	IT Solution
Physical resources		Members					Ledger
		Offers/Requests					Paper Tc
		Processes	**Processes**	**Processes**	**Processes**	**FB kind**	
			Coordinates (provides coordinator)	Matchmaking	Spending TC	+	
		Focus on TG(s)		SKART INFO receiving	Earning TC	+	
		SKART info		SKART INFO use for matchmaking	Work provided		
		Multiple TG SKART INFO	Builds rules and statute upon a law	Accessing the SKART database input	Work received		
	+ or -	Feedback about (ETQOR)		Accessing the SKART database read	Use of TC		
	+ or -	Info about O+R	Decision making	Events organizing	Issue of TC		
		Form interest groups		Networking	FB about transactions to coordination		
				Fundraising	Information about ETQOR		
				Managing			
				Security keeping			
		R: Wider system entities (Ownership level)		**Processes**			
		Citizens		Join			
		NGO		Founds, Funds, resources			
		Businesses		Founds, Funds, resources			
	+ or -	Government	Sub-Sub-System	Founds, Funds, resources, provides law frame			
			Governmental org.				
			Law				

Source: Author's research.

- Members can be from one or more focus groups; they form interest groups and they provide information about their SKART, and through this they create information about Offers and Requests, which feedbacks to the wider system. In addition, they provide feedback based on their own experience with exchanges, TC, quality and Offers and Requests.
- Base organization provides a framework by its statutes and rules, provides a coordinator and conducts decision making.
- Within or from outside a base organization, one or more coordinators is appointed. The coordinator performs matchmaking, receives SKART information from members, uses it for Matchmaking and uses the database of SKART to update information. Besides this, the coordinator provides networking and various events for members, takes cares of the funding of TB, manages it and keeps the security level high.
- Time credits are exchanged on the basis of an exchange platform, which is usually an IT solution, but it may also be a paper ledger or monetarized TC.
- Wider system entities include citizens, NGO, Businesses and Government and its organizations, all with roles outlined in the chart.
- Outputs are listed in following chart

The P2P model is the basic one in which people transform their SKART into offers and requests. All inputs are transformed to outputs listed in Box 2.

Person to Person Self-Managed Model

The Person to Person Self-Managed Model is shown in Figure 2 and Box 3.

Overview

Much of the information about this approach has been presented in different forms in the TB Taxonomy section as well as in other sections above; here it has been organized in terms of specific relevance regarding the P2P self-managed model, with the main differences marked in blue, i.e. basically emphasizing the fact that this model is based solely on people. It can be founded even only by a non-formal group of people or they can join in NGO which can found it. It does not require any coordinator, only an exchange platform. In theory this may be group of friends, colleagues or neighbors who decide to keep track of how they exchange between themselves and may just

Box 2. Outputs of P2P TB model

Outputs	
Impacts on "R" Entities	
Feedback of security keeping by coordinator and existence of rules of a TB raise credibility	
Information about TG variety stimulates joining	
Feedback about participation (motivation for recruit.)	
Feedback about spin-off events and activities of IG	
Savings	Social sphere
	Innovations
	Local business development
Systemic Outputs (on Outside Environment)	
Utilization of unused SKART	
Societal "bottom-up" problem solving by co-production (citizens use TB as a tool to solve own actual problems)	
Employability	
Non-formal Education	
Healthcare improvement (de-institutionalization)	
Awakening of dormant markets	
Reviving local economy	
Spin-off projects and innovation	
Learning - Enhanced Information and Knowledge sharing	
Systemic Feedbacks	
+	All outputs above induce positive feedback towards its own "R" entity
-	Misunderstanding about social/business role towards some "R" entities
-	Could be misunderstood by government and restricted/forbidden

Source: Author's research.

maintain a record in a paper ledger book, or they may decide to use one of free IT solutions for this. This is also the only model of TB which does not apply to businesses at all, although they might be affected if the TB grows in use. In practice, there may arise negative feedback in outputs related to lower security and the fact that there is no active stimulation of exchanges.

Person to Person Virtual Model of TB

The Person to Person Virtual Model of TB is shown in Figure 3 and Box 5.

Figure 2. Person to Person Self-Managed model of TB
Source: Author's research.

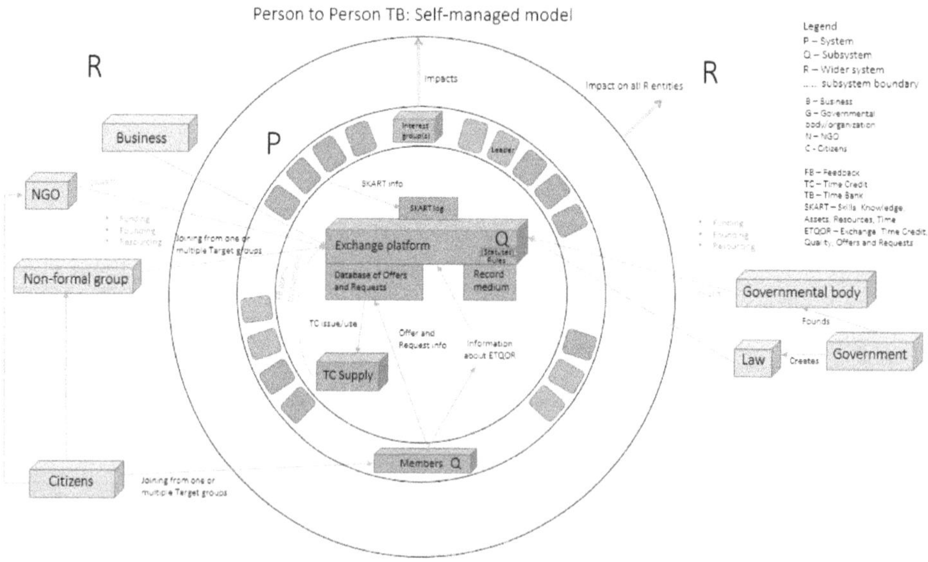

Overview

This is very similar in structure to the previous Self-managed scheme, i.e. also a very light version of TB. This approach is solely based on a virtual approach, so members might never meet in person. They are networked by an IT solution which is implemented by a person designated here as IT Solution admin, a person who may or may not be a member of the TB. As IT solutions might be free of charge, this model of TB comes at very low-cost or at no cost at all, which makes it very accessible. On the other hand, there are not many security issues aside of those involving feedback by the members (through which they can evaluate services) implemented into the IT solution. As security is very low, without a coordinator there is no one to match people with offers and requests or to keep and manage the TB itself. So, in summary, this model is very low-cost, but also a very unreliable approach. Outputs of virtual TB are listed in Box 6.

Person to Organization

Person to Organization is shown in Figure 4 and Box 7.

Box 3. Person to Person Self-Managed model of TB

Person to Person self-managed model of TB							
Inputs	**FB Kind**	**P: The system (Transformation process)**					
Funding only by NGO or Non-formal Group of citizens		**Q: Sub-Systems**					
		Membership	**Exchange platform instead of base organization**		**Time Credits**		
		Entities	**Entities**		**Entities**		
Founding		One TG					
Joining (people)		Multiple TG			Time Credit supply		
Recruitment (people)		Interest groups	*Exchange platform*			Sub-sub-system Q2	
Unused SKART		Members	SKART log			Record medium	IT Solution
SKART		Offers/Requests	Database of Offers and Requests				Ledger
Law		Leader(s) (non-formal coordinator)	Rules				Paper Tc
Physical resources							
		Processes	**Processes**		**Processes**	**FB kind**	
		Feedback about ETQOR only by Exchange platform			Spending TC	+	
					Earning TC	+	
				SKART INFO receiving	Work provided		
	+ or -	Info about O+R		SKART INFO use for matchmaking	Work received		
		Focus on TG(s)		Accessing the SKART database input	Use of TC		
		Form interest groups		Accessing the SKART database read	Issue of TC		
					FB about transactions to exchange **platform**		
					Offers/Requests		
					Information about ETQOR		
		R: Wider system entities (Ownership level)		**Processes**			
		Businesses		**Can be affected, can fund but usually are not founders**			
		Citizens		Join, resources, can fund by own funds			
		NGO		Founds, Funds, resources			
	+ or -	Government	Sub-Sub-System	Provides law frame			
			Governmental org.				
			Law				

Source: Author's research.

Box 4. Outputs of Person to Person Self-Managed model of TB

Outputs	
Impacts on "R" Entities (Note: in This Model Are Only C,N,G)	
Information about TG variety stimulates joining	
Feedback about participation (motivation for recruit.)	
Feedback about spin-off events and activities of IG	
Savings	Social sphere
	Innovations
Systemic Outputs (on Outside Environment)	
Utilization of unused SKART	
Societal "bottom-up" problem solving by co-production (citizens use TB as a tool to solve own actual problems)	
Employability	
Non-formal Education	
Healthcare improvement (de-institutionalization)	
Awakening of dormant markets	
Reviving local economy	
Spin-off projects and innovation	
Active citizenship	
Learning - Enhanced Information and Knowledge sharing	
Systemic Feedbacks	
+	All outputs above induce positive feedback towards its own "R" entity
-	Misunderstanding about social/business role towards some "R" entities
-	Could be misunderstood by government and restricted/forbidden
-	Lower security
-	No active stimulation of offer and requests

Source: Author's research.

Overview

A Person to Organization TB is founded with a specific goal, usually to solve some problem within a region. It operates on the principle that a base organization is a center or a bank which provides activity offers and rewards. Compared to the possible alternative, that people will perform an offered job for a reward, it has the advantage that it may entail more activities and more rewards, and thus the TC can be spent on various rewards over time. The difference from giving members money is that the either the activities or the rewards might not be available in the labor or other markets. This

Figure 3. Person to Person virtual model of TB
Source: Author's research.

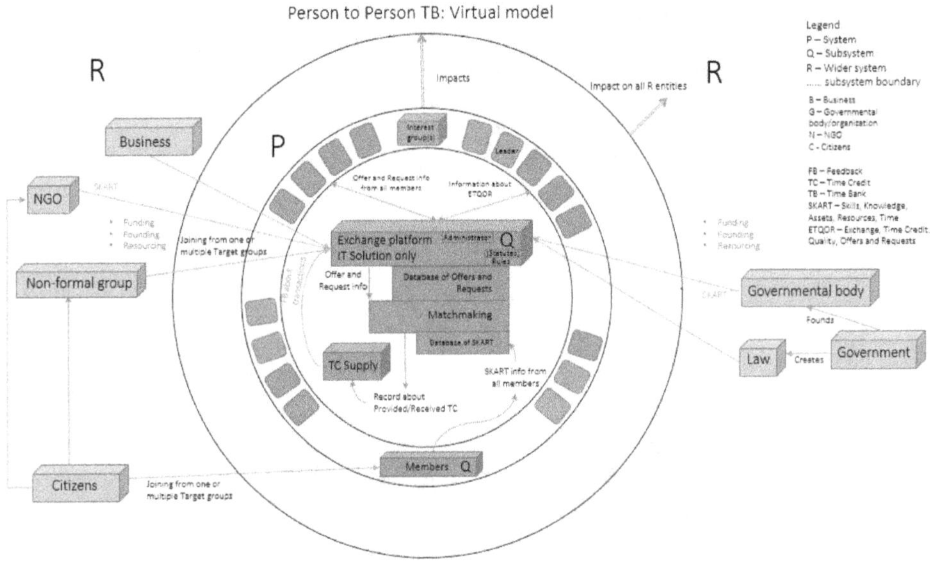

approach is used mostly by governmental organizations, who acknowledge a particular problem in the region and would like to solve it by co-production. This does not just consist of investing money, but actively involving people in problem solving. The secondary impact of TB is the opportunity to form interest groups, as indicated above, and, with this, spin-off activities. In addition, the P2O variant is often a starter initiative of TB installed in a "top down" approach. The main differences visible in the chart are that the TB is founded with a mission (specific goal) and in the above-mentioned fact that there are activities proposed and rewards to use the TC for. The rewards and activities may also be proposed by other organizations including businesses, which are here outlined as "Partner organizations." When a member performs a designated task, he or/she "logs" the work into the record medium of the exchange platform, and it is within the exchange platform later approved or not by the coordinator. Outputs of this model of TB are shown in Box 8.

Person to Organization: Homogenous

Homogenous Person to Organization is shown in Figure 5 and Box 9.

Box 5. Person to Person virtual model of TB

Person to Person Virtual model of TB

Inputs	FB Kind	P: The system (Transformation process)					
Funding		Q: Sub-Systems					
Founding		**Membership**	**Exchange platform instead of base organization**		**Time Credits**		
Joining (people)		**Entities**	**Entities**		**Entities**		
Recruitment (people)		One TG	Matchmaking element in the IT Solution		Time Credit supply		
Unused SKART		Multiple TG					
SKART		Interest groups	*Exchange platform is strictly an IT Solution*			Sub-sub-system Q2	
Law		Members	Database of SKART			Record medium	IT Solution
Physical resources		Offers/Requests	Database of Offers and Requests				
IT Solution admin		Leader(s) (non-formal coordinator)	Rules				
							Administrator
		Processes	**Processes**		**Processes**	**FB kind**	
		Feedback about ETQOR only by Exchange platform	Matchmaking		Spending TC	+	
					Earning TC	+	
				SKART INFO receiving	Work provided		
		Focus on TG(s)		SKART INFO use for matchmaking	Work received		
		SKART info		Accessing the SKART database input	Use of TC		
	+ or -	Info about O+R		Accessing the SKART database read	Issue of TC		
		Form interest groups			FB about transactions to exchange platform		
					Offers/Requests		
					Information about ETQOR		
		R: Wider system entities (Ownership level)		**Processes**			
		Citizens		Join			
		NGO		Founds, Funds, resources			
		Businesses		Founds, Funds, resources			
	+ or -	Government	Sub-Sub-System	Founds, Funds, resources, provides law frame			
			Governmental org.				
			Law				

Source: Author's research.

Box 6. Outputs of Person to Person virtual model of TB

Outputs	
Impacts on "R" Entities	
Information about TG variety stimulates joining	
Feedback about participation (motivation for recruit.)	
Feedback about spin-off events and activities of IG	
Simple to maintain, low-cost to free	
Savings	Social sphere
	Innovations
	Local business development
	Autonomy and low cost
Systemic Outputs (on Outside Environment)	
Utilization of unused SKART	
Societal "bottom-up" problem solving by co-production (citizens use TB as a tool to solve own actual problems)	
Employability	
Non-formal Education	
Healthcare improvement (de-institutionalization)	
Awakening of dormant markets	
Reviving local economy	
Spin-off projects and innovation	
Active citizenship	
Learning - Enhanced Information and Knowledge sharing	
Systemic Feedbacks	
+	All outputs above induce positive feedback towards its own "R" entity
-	Misunderstanding about social/business role towards some "R" entities
-	Could be misunderstood by government and restricted/forbidden
-	Lower security
-	No active stimulation of offer and requests

Source: Author's research.

Overview

The homogenous model of P2O has only one adaptation, i.e. it focuses on only one service provided, which also usually, but not necessarily, limits it to one target group. Thus, the system is almost the same as the previous one, but the difference is that we can call the general P2O approach a strategic version, and the homogenous variant a tactical one. Homogenous P2O is

Figure 4. Person to organization pure model of TB
Source: Author's research.

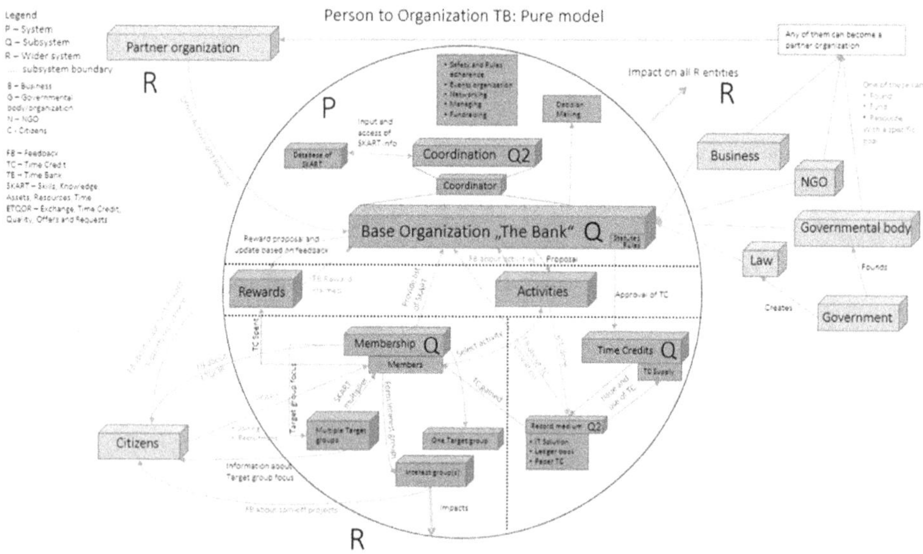

aimed at one object and one goal, and this objective might change over time; or the TB may transform into another arrangement; or it may simply fulfil its mission and be closed. Outputs are very similar and are listed in Box 10.

Person to Organization: Pension Scheme

Pension Scheme Person to Organization is shown in Figure 6 and Box 11.

Overview

The pension scheme is a very specific variant of the P2O approach, with not many having as of yet come into existence. Nevertheless, to provide a complete picture it is necessary to analyze the model carefully. This model encompasses slight changes in how the work is done. Members are generally divided into two groups, Givers and Takers. Givers are at standard active working age and can invest their unused SKART as well as offer services to others, but rather than spending TC straight out, they "save" them for later. By this they are accumulating TC until a situation arises in which they would need them, e.g. a momentary crisis. The idea, however, is to keep them until

Box 7. Person to organization pure model of TB

Person to Organization pure model of TB (Organization - The Bank, offers services and rewards for them; founded with a specific goal)							
Inputs	**FB kind**	**P: The system (Transformation process)**					
Funding		**Q: Sub-Systems**					
Founding with a specific goal		**Membership**	**Base Organization**		**Time Credits**		
		Entities	**Coordination (is Q within base org)**		**Entities**		
Joining		One TG	**Entities**	**Coordination Q2**	Time Credit supply		
Recruitment (people)		Multiple TG	Statutes	**Entities**	Sub-sub-system Q2		
Unused SKART		Interest groups	Rules	Coordinator	Record medium		IT Solution
SKART		Members		Database of SKART			Ledger
Law							Paper Tc
Physical resources							
Rewards offer							
Activities offer	+ or -		Proposed Rewards (I)				
		Processes	**Processes**	**Processes**	**Processes**	**FB kind**	
		Select activity	Builds rules and statute upon a law	Matchmaking	Spending TC	+	
		SKART info		SKART INFO receiving	Earning TC	+	
		Multiple TG SKART INFO	Coordinates (provides coordinator)	SKART INFO use for matchmaking	Work provided		
	+ or -	Feedback about (ETQOR)		Accessing the SKART database input	Work received		
	+ or -	Fb from M about rewards usability	Decision making	Accessing the SKART database read	Use of TC		
		Focus on TG(s)	Activities proposal	Events organizing	Issue of TC		
		Form interest groups		Networking	FB about transactions to coordination		
				Security keeping	Approval of TC		
				Fundraising			
				Managing			
		R: Wider system entities (Ownership level)		**Processes**			
		Citizens		Join			
		NGO		Founds, Funds, resources			
		Businesses		Founds, Funds, resources			
	+ or -	Government	Sub-Sub-System	Founds, Funds, resources, provides law frame			
			Governmental org.				
			Law				
		Partner organization		Can offer activities and propose rewards via Base org			

Source: Author's research.

Box 8. Output of Person to organization pure model of TB

Outputs	
Impacts on "R" Entities	
Recruiting activities by coordinator/stimulation to join by a reward and/or activity prop.	
Information about TG variety stimulates joining	
Feedback about participation (motivation for recruit.)	
Feedback about spin-off events and activities of IG	
Proposed rewards are usually of lower value than benefit done by an activity	
Savings	Social sphere
	Innovations
	Local business development
Systemic Outputs (on Outside Environment)	
Utilization of unused SKART	
Societal problem solving by co-production (citizens use TB as a tool to solve problems) but it is installed from "top down"	
Employability	
Non-formal Education	
Healthcare improvement (de-institutionalization)	
Awakening of dormant markets	
Reviving local economy	
Spin-off projects and innovation	
Learning - Enhanced Information and Knowledge sharing	
Systemic Feedbacks	
+	All outputs above induce positive feedback towards its own "R" entity
-	Misunderstanding about social/business role towards some "R" entities
-	Could be misunderstood by government and restricted/forbidden
-	"Top down" approach can lack interest of target groups

Source: Author's research.

they are of pension age, when the TC can be used to gain other services they might need at that time.

Organization to Organization Model

The Organization to Organization Model is shown in Figure 7 and Box 13.

Figure 5. Person to organization TB: Homogenous
Source: Author's research.

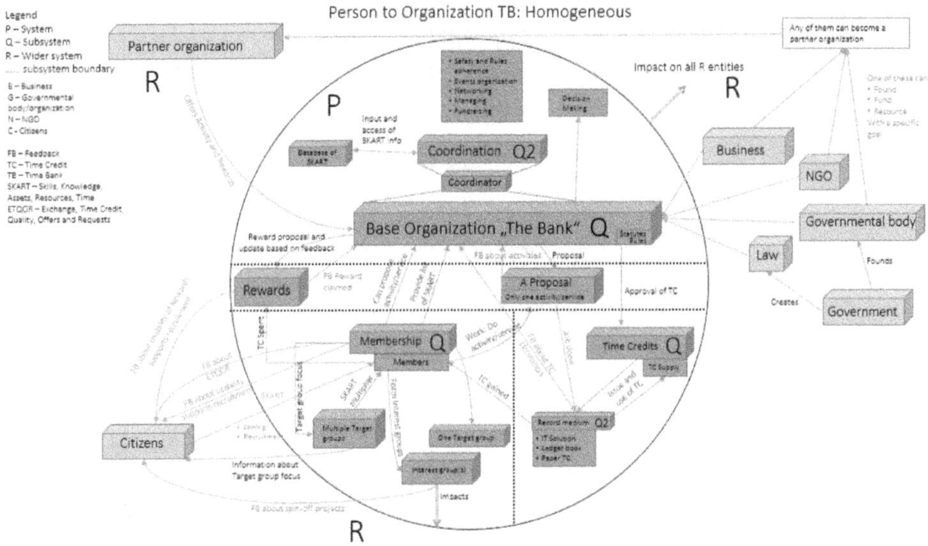

Overview

Organization to organization is a very advanced variant of TB. It incorporates only organizations who see an advantage in networking to facilitate, e.g. the sharing and exchange of resources which are not immediately in use, and, by this, optimize their overall efficiency. The composition of organizations in the TB can vary greatly, e.g. it may contain mainly NGOs, businesses and also governmental organizations. Geographically, this kind of TB is usually concentrated in one region and the main aim is usually to develop the region as a whole by developing in systemic way all the organizations involved through the sharing of common resources. The cooperation comes from the understanding that some socially problematic situations which NGOs and local government are dealing with can be resolved with the help of businesses, and vice versa, i.e. that healthy business in the region results from having fewer social problems. It does not matter which organization is the founder; it can be any of parties involved. In this model citizens are the entity which is affected solely by the outputs of the TB and they participate only indirectly as part of organizations. As indicated above, entrepreneurs, which are physical bodies, can also be counted among organizations if they enter as a business entity. As for the exchange platform, on this scale any information platform other than IT Solutions would not be practical.

Box 9. Person to organization Homogenous model of TB

Person to Organization homogenous model of TB (Homogenous mean that only one service/activity is proposed to members for designated rewards)							
Inputs	**FB Kind**	**P: The system (Transformation process)**					
Funding		**Q: Sub-Systems**					
Founding with a specific goal		**Membership** **Entities**	**Base Organization** **Coordination (is Q within base org)**		**Time Credits** **Entities**		
Joining		One TG	**Entities**	**Coordination Q2**	Time Credit supply		
Recruitment (people)		Multiple TG	Statutes	**Entities**	Sub-sub-system Q2		
Unused SKART		Interest groups	Rules	Coordinator	Record medium		IT Solution
SKART		Members					Ledger
Law							Paper Tc
Physical resources							
Rewards offer							
Activity offer							
	+ or -		Proposed Rewards				
		Processes	**Processes**	**Processes**	**Processes**	**FB kind**	
		Select the one activity	Builds rules and statute upon a law	Matchmaking	Spending TC	+	
		Focus on TG(s)		Events organizing	Earning TC	+	
			Coordinates (provides coordinator)	Networking	Work provided		
	+ or -	Feedback about (ETQOR)		Fundraising	Work received		
	+ or -	Fb from members about rewards usability	Decision making	Managing	Use of TC		
		Form interest groups	Activity proposal	Security keeping	Issue of TC		
		SKART info			Approval of TC		
		Multiple TG SKART INFO			FB about transactions to coordination		
		R: Wider system entities (Ownership level)		**Processes**			
		Citizens		Join			
		NGO		Founds, Funds, resources, can become a partner org		Work proposal and rewards can come from any	
		Businesses		Founds, Funds, resources, can become a partner org			
	+ or -	Government	Sub-Sub-System	Founds, Funds, resources, provides law frame, can become a partner org			
			Governmental org.				
			Law				
		Partner organization		Can offer activity and propose rewards via Base org			

Source: Author's research.

Figure 6. Person to Organization TB Pension scheme
Source: Author's research.

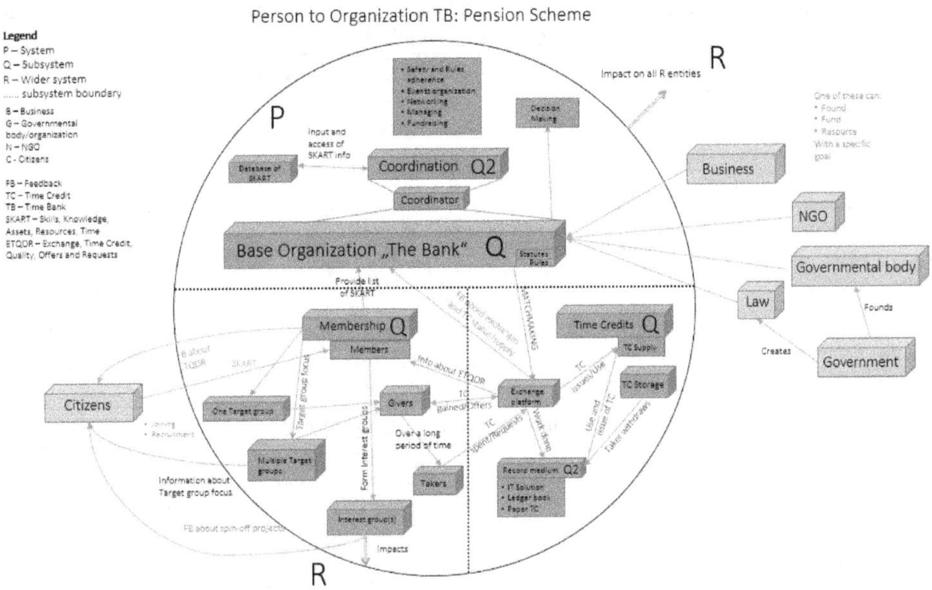

Organization to Organization Plus People

The Organization to Organization Plus People model is shown in Figure 8 and Box 15.

Overview

Organization to Organization plus people is probably the most advanced version of a single TB which exists. It uses a powerful sharing template across sectors from O2O and adds citizens and entices them to share their own SKART. By this connection, sharing and exchange opportunities multiply and most of free resources in a region can be offered, requested and used. In addition, such a wide variety of entities and target groups offers various points of view regarding various problems and this interaction is usually the source of many spin-off and innovative projects. If a TB develops to this scope and relatively large size (usually quite extensive in terms of numbers of members when all possible entities are involved) it is able to transform relative small inputs into strong outputs. Outputs are again listed in Box 16.

Box 10. Outputs of Person to organization TB: Homogenous

Outputs	
Impacts on "R" Entities	
Recruiting activities by coordinator/stimulation to join by a reward and/or activity prop.	
Information about TG variety stimulates joining	
Feedback about participation (motivation for recruit. By a reward)	
Feedback about spin-off events and activities of IG	
Focus on solving of a specific problem important for an Organization (entity)	
Proposed rewards are usually of lower value than benefit done by an activity	
Savings	Social sphere
	Innovations
	Local business development
Systemic Outputs (on Outside Environment)	
Utilization of unused SKART (could be just basis for "can or cannot do it" selection of prop.activity)	
Societal problem solving by co-production (citizens use TB as a tool to solve problems) but it is installed from "top down" by base org	
Employability	etc. Depends on needs of base organization or/and its founding entity
Non-formal Education	
Healthcare improvement (de-institutionalization)	
Awakening of dormant markets	
Reviving local economy	
Spin-off projects and innovation	
Learning - Enhanced Information and Knowledge sharing	
Systemic Feedbacks	
+	All outputs above induce positive feedback towards its own "R" entity
-	Misunderstanding about social/business role towards some "R" entities
+	Often founded by a governmental body to solve specific problems
-	"top down" approach can lack interest of target groups

Source: Author's research.

Time Bank Networks

Although TB networks are not very common, where they do exist they are

Box 11. Person to Organization TB Pension scheme model

Person to Organization approach - Pension scheme (specific model of TB designed to "Store" TC after being earned, to use them later over long period of time as a pension supplement)							
Inputs	FB kind	P: The system (Transformation process)					
Funding		Q: Sub-Systems					
Founding		Membership	Base Organization		Time Credits		
Joining		Entities	Coordination (is Q within base org)		Entities		
Recruitment (people)		One TG	Entities	Coordination Q2	Time Credit supply		
Unused SKART		Multiple TG	Statutes	Entities	Exchange platform	Sub-sub-system Q2	
SKART		Interest groups	Rules	Coordinator		TC storage	IT Solution
Law		Givers		Database of SKART			Ledger
Physical resources		Takers					Paper Tc
						Offers (givers)	
						Requests (takers)	
	+ or -						
		Processes	Processes	Processes	Processes	FB kind	
	+	Giver change to Taker over long time	Builds rules and statute upon a law	Matchmaking	Spending TC (taker)	+	
		Focus on TG(s)		SKART INFO receiving	Earning TC (giver)	+	
				SKART INFO use for matchmaking	Work provided		
	+ or -	Feedback about (ETQOR)	Coordinates (provides coordinator)	Accessing the SKART database input	Work received		
	+ or -	Info about O+R		Accessing the SKART database read	Use of TC		
		SKART info	Decision making	Events organizing	Issue of TC		
		Multiple TG SKART INFO		Networking	FB about transactions to coordination		
		Form interest groups		Security keeping	Giver save		
				Fundraising	Taker withdraws		
				Managing	Information about ETQOR		
		R: Wider system entities (Ownership level)		Processes			
		Citizens		Join (needs legal entity so founded by citizens only through an organization)			
		NGO		Founds, Funds, resources, can become a partner org			
		Businesses		Founds, Funds, resources, can become a partner org			
	+ or -	Government	Sub-Sub-System	Founds, Funds, resources, provides law frame			
			Governmental org.				
			Law				

Source: Author's research.

characterized by common governance, at least partially, to be able to focus on common goals. This usually broadens the possibilities of exchange, as many of the IT Solutions offer transferability of TC, which also opens up new possibilities of exchange outside of region. Also, networks aim for higher visibility of transactions and greater advocacy for various stakeholders. A very short overview is shown in Box 17.

OUTPUTS SUMMARIZED FROM ALL NINE MODELS

In the following lines a complete summary of outputs of all types of TB will be introduced. The outputs have been grouped topically, as they form similar groupings even when they proceed from different TB models. The last important point in this sub-chapter is a list of systemic feedbacks.

Box 12. Outputs of Person to Organization TB Pension scheme

Outputs		
Impacts on "R" Entities		
Recruiting activities by coordinator/stimulation to join		
Information about TG variety stimulates joining (I)		
Feedback about participation (motivation for recruitment) (I) + (K)		
Feedback about spin-off events and activities of IG (I) + (K)		
Focus on solving of a specific problem important for a society (entity all involved entities)		
Feedback of security keeping by coordinator and existence of rules of a TB raise credibility		
Savings	Social sphere	Pensions
	Innovations	
	Local business development	
Systemic Outputs (on Outside Environment)		
Utilization of unused SKART (to "save" them for later and use when needed, in retirement		
Societal problem solving by co-production (citizens bring impact and as a by-product they are saving for future)		
Employability		
Non-formal Education		
Healthcare improvement (de-institutionalization), especially important here, aimed to solve issues of elderly)		
Awakening of dormant markets (elderly)		
Reviving local economy		
Spin-off projects and innovation		
Learning - Enhanced Information and Knowledge sharing		
Systemic Feedbacks		
+	All outputs above induce positive feedback towards its own "R" entity	
+	By awakening SKART in elderly new area of economy becomes active, which also brings savings	
+	Often founded by a governmental body to solve specific problems	
+	Can work as a regular TB before the service savings (TC) build up	
-	Misunderstanding about social/business role towards some "R" entities	
-	Comes from a need of a region	
-	Requires long time for services build up	

Source: Author's research.

The coordinator (with all duties seen in processes of coordination), amount of SKART (which is related to the amount of members), rewards (in certain scenarios) and activities all have a direct impact on the recruitment of new members (both people and organizations).

Figure 7. Organization to organization TB model
Source: Author's research.

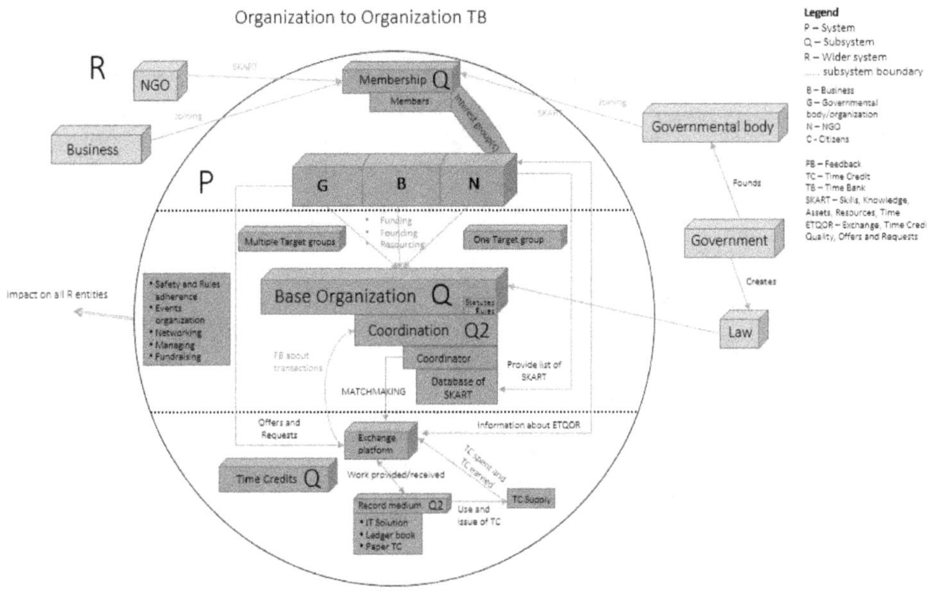

Aside from the focus of the whole TB on the target group(s), it is important to focus on solving specific problems determined by the organization which founded the TB (whether it is a NGO, Business, Governmental organization, or even a non-formal group of people), or from a holistic point of view, to focus on a societal problem.

Feedbacks which flow outside of the transformation area of the system (P) demonstrate how a TB works. With effective coordination and rules, security is higher, which might motivate more members to join. When members are active and form interest groups, they also produce spin-off activities, which have a direct impact in the area on which they are aimed. And the greater the variety of SKART involved, the more efficient a TB becomes.

This identifies the flow of information from Members to R (wider system) entities regarding the status of a variety of target groups, so more target groups can join.

Savings present one of the most important outputs. TB is a very low-cost tool to facilitate all of the outputs above as well as the impacts mentioned below. This is accomplished by giving people, and on a larger scale organizations, including businesses (especially local ones), a tool to facilitate exchange and sharing of resources they have, but cannot use. This flow of offers and requests

Figure 8. Organization to organization plus People TB model
Source: Author's research.

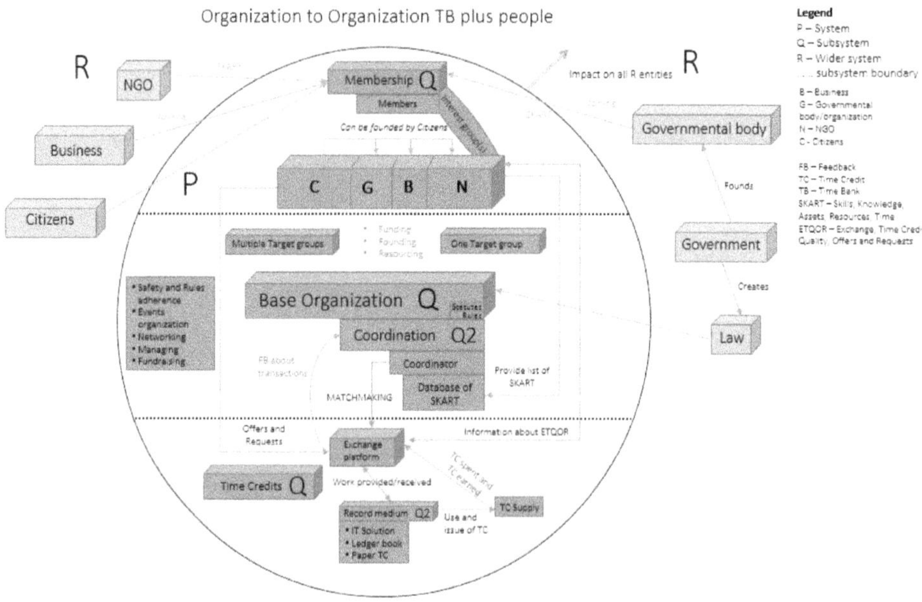

can save a lot of money, or it can help solve financial problems. A government would thus not need to invest into so many various types of social schemes. Companies can utilize unused resources and thus make business run more efficiently. By bringing various target groups together, fields become open to innovation, an outcome which does not require high financial stimuli. By its very nature TB requires only limited inputs. Insignificant funds produce potentially massive outputs. Everything depends on the will of government and stakeholders, and, more potentially powerful, on people and their SKART.

The utilization of unused SKART is a central feature of any TB, starting from the decision as to whether SKART is enough for a proposed task, such as in P2O - Homogenous, to proposals of SKART between all wider system entities in the O2OP model. The strong side of TB is in its possibility to transform almost any SKART into an offer and/or request, and to find another member which would use the potential. All of the bodies, physical or legal, have SKART which can be brought to use, e.g. as a by-product, in any economic situation or society, including all the impacts mentioned above and below.

TB are most often founded on a basis of the need to solve a certain problem – whether with the local economy and the activity of citizens along with the unused potential of either, with problems of local government, or with the

Box 13. Organization to organization pure TB model

Organization to Organization pure model of TB								
Inputs	**FB Kind**	**P: The system (Transformation process)**						
Funding		**Q: Sub-Systems**						
Founding		**Membership**		**Base Organization**		**Time Credits**		
Joining		**Entities**		**Entities**				
Recruitment (organizations)		Members are organizations from R	NGO	**Coordination (is Q within base org)**		**Entities**		
			Business	**Entities**	**Coordination Q2**	Time Credit supplly		
Unused SKART			Governmental org.	Statutes	**Entities**	Exchange platform	Sub-sub-system Q2	
SKART		Any member organization can become Base organization		Rules	Coordinator		Record medium	IT Solution
Law		Offers/Requests			Database of SKART			Other non-practical
Physical resources								
		Processes		**Processes**	**Processes**	**Processes**	**FB kind**	
		Form interest groups		Coordinates (provides coordinator)	Matchmaking	Spending TC	+	
		Focus on TG(s)		Builds rules and statute upon a law	SKART INFO recieving	Earning TC	+	
		SKART info			SKART INFO use for matchmaking	Work provided		
		Multiple TG SKART INFO		Decision making	Accessing the SKART database input	Work recieved		
	+ or -	Feedback about (ETQOR)			Accessing the SKART database read	Use of TC		
	+ or -	Info about O+R			Events organizing	Issue of TC		
					Networking	FB about transactions to coordination		
					Fundraising			
					Managing	Information about ETQOR		
					Security keeping			
		R: Wider system entities (Ownership level)			**Processes**			
		Citizens			Might be just affected by outputs			
		NGO			Founds, Funds, resources			
		Businesses			Founds, Funds, resources			
	+ or -	Government		Sub-Sub-System	Founds, Funds, resources, provides law frame			
				Governmental org.				
				Law				

Source: Author's research.

Box 14. Outputs of Organization to organization TB model

Outputs	
Impacts on "R" Entities	
Recruiting activities by coordinator to raise amount of services and SKART involved	
Free unused SKART can be used to enhance systemic impact	
Feedback about participation (motivation for recruit.)	
Feedback about spin-off events and activities of IG	
Feedback of security keeping by coordinator and existence of rules of TB raise credibility	
Savings	Social sphere
	Innovations
	Local business development
Systemic Outputs (on Outside Environment)	
Utilization of unused SKART (both receiving and providing); Organizational free capacities can be valuable to offer, but also to receive, but without TB they stay unused	
Regional problem solving	
Employability	
Non-formal Education	
Business co-production (taking advantage of innovative approaches to share and develop)	
Awakening of dormant markets	
Reviving local economy	
Spin-off projects and innovation	
Learning - Enhanced Information and Knowledge sharing	
Systemic Feedbacks	
+	Supports activity of organizational operation (by co-production new markets and fields of operation can be uncovered
+	All outputs above induce positive feedback towards its own "R" entity
+	Unused resources by some organizations are used by ones who require them
-	Misunderstanding about social/business role towards some "R" entities
-	Could be misunderstood by government and restricted/forbidden

Source: Author's research.

necessity of transferring skills of knowledge. In any of these cases TB can be used as a tool. The two main approaches mentioned in terms of impact are "bottom up" and "top down." The first has the advantage of naturally forming a TB around a problem. Members (no matter whether people or organizations) themselves gather to resolve the issue(s) and through this process possible spin-off projects and innovations appear. The latter comes usually from an organization, with TB "installed" from the top recruiting member entities to solve a problem.

Box 15. Organization to organization plus People TB model

Organization to Organization pure model of TB - hybrid model including also People								
Inputs	**FB kind**	**P: The system (Transformation process)**						
Funding		**Q: Sub-Systems**						
Founding		**Membership**		**Base Organization**				
Joining		**Entities**		**Entities**				
Recruitment (organizations and people)		Members are entities from "R"	NGO	**Coordination (is Q within base org)**		**Entities**		
			Business	**Entities**	**Coordination Q2**	Time Credit supply		
			Governmental org.	Statutes	**Entities**	Exchange platform	Sub-sub-system Q2	
Unused SKART			Citizens can also join	Rules	Coordinator		Record medium	IT Solution
SKART		Any member organization can become or found a Base organization			Database of SKART			Other non-practical
Law								
Physical resources		Offers/Requests						
		Processes		**Processes**	**Processes**	**Processes**	**FB kind**	
		Form interest groups		Coordinates	Matchmaking	Spending TC	+	
		Focus on TG(s)		Builds rules and statute upon a law	SKART INFO receiving	Earning TC	+	
		SKART info			SKART INFO use for matchmaking	Work provided		
		Multiple TG SKART INFO		Decision making	Accessing the SKART database input	Work received		
	+ or -	Feedback about (ETQOR)			Accessing the SKART database read	Use of TC		
	+ or -	Info about O+R			Events organizing	Issue of TC		
					Networking	FB about transactions to coordination		
					Fundraising			
					Managing	Information about ETQOR		
					Security keeping			
		R: Wider system entities (Ownership level)			**Processes**			
		Citizens			Can join, or join/create a founding organization			
		NGO			Founds, Funds, resources			
		Businesses			Founds, Funds, resources			
	+ or -	Government		Sub-Sub-System	Founds, Funds, resources, provides law frame			
				Governmental org.				
				Law				

Source: Author's research.

Box 16. Outputs of Organization to organization plus People TB model

Outputs	
Impacts on "R" Entities	
Recruiting activities by coordinator to raise amount of services and SKART involved	
Feedback about Variety of SKART raises impact on participation on TB	
Feedback about participation (motivation for recruit.)	
Feedback about spin-off events and activities of IG	
Feedback of security keeping by coordinator and existence of rules of TB raise credibility	
Savings	Social sphere
	Innovations
	Local business development
Systemic Outputs (on Outside Environment)	
Utilization of unused SKART by organizations (both receiving and providing); Organizational free capacities can be valuable to offer, but also to receive, but without TB they stay unused	
Utilization of unused SKART by citizens (free resources used)	
Free unused SKART can be used to enhance systemic impact	
Societal "bottom-up" problem solving by co-production (citizens use TB as a tool to solve own actual problems)	
Employability	
Non-formal Education	
Business co-production (taking advantage of innovative approaches to share and develop)	
Awakening of dormant markets	
Reviving local economy	
Spin-off projects and innovation	
Learning - Enhanced Information and Knowledge sharing	
Healthcare improvement (de-institutionalization)	
All actors (not only R entities, but people within) can share skills and knowledge	
Regional problem solving	
Systemic Feedbacks	
+	Supports activity of organizational operation (by co-production new markets and fields of operation can be uncovered
+	All outputs above induce positive feedback towards its own "R" entity
+	Unused resources by some organizations are used by ones who need them
-	Misunderstanding about social/business role towards some "R" entities
-	Could be misunderstood by government and restricted/forbidden

Source: Author's research.

Box 17. Time Bank networks

Input		
Whole different TBs Join with their members, target groups, interest groups, SKART etc. (All kinds of previous models)	Members	Exchange Platform provides coordination
	Entities	Processes
	Various TBs connected to exchange platform	Matchmaking
		Events organizing and Networking
	A decision making sub-system Q2	Safety and Rules adherence
	A decision making body	Fundraising
		Managing
	Processes	Offer and Request sharing
	Exchanging information about O+R	Record of exchanges
	Exchanging information about SKART	Record of TC flow
	Decision making	ETQOR Feedback
Output		
All outputs of previous models are multiplied by possibility to exchange more services, and more inputs by SKART and more variety in general.		

Source: Author's research.

Table 1.

Recruitment	
	Feedback about participation (motivation for recruit.) (I) + (K)
	Recruiting activities by coordinator to raise amount of services and SKART involved
	Recruiting activities by coordinator/stimulation to join by a reward and/or activity prop.

Table 2.

Strengthened focus	
	Focus on solving of a specific problem important for an Organization (entity)
	Focus on solving of a specific problem important for a society (entity all involved entities)

Table 3.

Feedbacks	
	Feedback of security keeping by coordinator and existence of rules of a TB raise credibility
	Feedback about spin-off events and activities of IG (I) + (K)
	Feedback about Variety of SKART raises impact on participation on TB

Table 4.

Information	Information about TG variety stimulates joining (I)

Table 5.

Savings	Proposed rewards are usually of lower value than benefit done by an activity					
	Simple to maintain, low-cost to free					
	Savings in/for					
	Governmental					
	Social sphere					
	Local business development					
	Innovations					
	Autonomy and low cost					

Table 6.

SKART	Utilization of unused SKART
	Utilization of unused SKART (could be just basis for "can or cannot do it" selection of prop. activity)
	Utilization of unused SKART (to "save" them for later and use when needed, in retirement)
	Utilization of unused SKART (both receiving and providing); Organizational free capacities can be valuable to offer, but also to receive, but without TB they stay unused
	Utilization of unused SKART (both receiving and providing); Organizational free capacities can be valuable to offer, but also to receive, but without TB they stay unused
	Free unused SKART can be used to enhance systemic impact
	Utilization of unused SKART by citizens (free resources used)

These outputs are the by-products of the ones mentioned above: simply the result of exchanges occurring, even if these only take the form of learning and exchanging information and knowledge or other educational activities. As most activities are work based, employability rises with the higher skills of members. Healthcare improvement comes from solving the non-medical based problems which are now dealt with by physicians, but are not directly related to health (i.e. problems caused by "civilization" illnesses, which can be more effectively solved by means other than medication). By involving

Table 7.

Problem solving	Business co-production (taking advantage of innovative approaches to share and develop)
	All actors (not only R entities, but people within) can share skills and knowledge
	Regional problem solving
	Societal "bottom-up" problem solving by co-production (citizens use TB as a tool to solve own actual problems)
	Societal problem solving by co-production (citizens use TB as a tool to solve problems) but it is installed from "top down" by base org./or other stakeholders
	Societal problem solving by co-production (citizens bring impact and as a by-product they are saving for future)
	Societal "bottom-up" problem solving by co-production (citizens use TB as a tool to solve own actual problems)

Table 8.

Other Impacts on environment	Employability
	Non-formal Education
	Healthcare improvement (de-institutionalization)
	Awakening of dormant markets
	Reviving local economy
	Spin-off projects and innovation
	Learning - Enhanced Information and Knowledge sharing
	Active citizenship

Table 9. Systemic feedbacks

+	All outputs above induce positive feedback towards its own "R" entity
+	Often founded by a governmental body to solve specific problems
+	Can work as a regular TB before the service savings (TC) build up
+	By awakening SKART in elderly new area of economy becomes active, which also brings savings
+	Unused resources by some organizations are used by ones who need them
-	Comes from a need of a region
-	Misunderstanding about social/business role towards some "R" entities
-	Could be misunderstood by government and restricted/forbidden
-	Lower security
-	No active stimulation of offer and requests
-	Comes from a need of a region
-	Requires long time for services build up

Source: Author's research.

the business element, but also by basic sharing between members, the local economy can become more productive. And by the ability of TB to offer/receive any new SKART, innovation, spin-off projects are uncovered, all of which leads to discovery of new markets as well as new possible products and services which can be also opened up to the market.

SYSTEMIC FEEDBACKS

Systemic feedbacks are shown in Table 9.

Positive Feedbacks

- All outputs related to a wider system entity will have positive feedbacks. This means positive feedback regarding the founding organization, as this entity is the owner (i.e. the ownership level in the wider system according to CATWOE (Checkland, 1999)) of the TB and it has power to start, open or close it. In other words, a TB is founded by a R entity for a reason, and this reason brings desirable outputs. If these benefits do not come, there would be a reason to close the TB. On the other hand, this is actually the most valid positive feedback, because it also means that all the outputs listed above provide positive feedback to their own R entities which allow their inputs into a TB system to be transformed into outputs.
- Related to the above, very often a TB is founded by a governmental body to solve a specific problem. This in turn creates other positive feedback loops, as with increased problem-solving activity the governmental body receives acclaim, and with a governmental body as a founding institution the TB is established and operates with higher degree of credibility, and finally the positive feedback flows outward to attract other potential members.
- Related to pension scheme TB, this model can provide the positive feedback as a regular TB, which would in turn actually would bring more results besides the standard benefits of the pension scheme sub-version.
- This feature is related to specific feedback in utilizing SKART in the elderly population by their inclusion into TB. Benefits include the various offers enabled by seniors by their considerable knowledge and

experience on one side, along with their very specific requests, which are usually services unavailable on market, not to mention savings in healthcare for governmental institutions.

- Resources of organizations which are now unused can be utilized and counter value can be gained to optimize the efficiency of their operations. Even without financial resources, new possibilities open up to facilitate gains in other kinds of resources and services.
- Most often, founding a TB is not a random event. It proceeds from a particular need in a region and TB is considered to be a tool for problem solving with a chance to bring a broader positive impact on regional development.

Negative Feedbacks

- The role of TB might be misunderstood by some entities. Businesses in particular often consider TB to be merely a social system, or worse, a heretical attack on the economic system, and thus instead of obtaining benefits from participation they condemn it from the beginning.
- The same situation can occur on a governmental level. Even though there are many countries, regions (actual and virtual) and municipalities where TB works legally and prosperously alongside of the market economy a blindfolded view of this socio-economic innovation may hinder all the types of development made possible only by TBs.
- Lower security is a feedback which is present only in the P2P Self-managed and Virtual models, as these do not have a coordinator which would oversee security issues.
- No active stimulation represents the same situation as above, i.e. the same models and the fact that there is no coordinator present.
- The requirement of a long time period for services to build up is related to the pension scheme P2O model, where this fact can cause negative feedback "outside" a TB.

The analysis of rich pictures which were developed into diagrams and charts attempted to display all the basic elements of TBs by breaking down each of approaches as seen in the real world. Now the way is open to use these elements and to construct a universal model.

As the building blocks of TB, these elements could conceivably grow in number if new approaches to TB appear in the world, so the number of possibilities presented here is not final. But even these elements explained above are enough to construct a new model, not only a universal one, but also one optimized and without redundancies. The elements also can be taken into consideration, and a new approach can be developed using the universal TB model as a template.

ROOT DEFINITIONS

In this part, root definitions are formulated based on the results of the analysis. The main verbs which they contain as they are related to models above are: *transform, solve, exchange, offer, request, use SKART*

P2P

A system which connects people and transforms their unused skills, knowledge, assets, resources and time into offers and requests by creating a network coordinated and secured by a coordinator through which people who become members. Members share their offers and requests for services useful for them. These offers and requests are exchanged equally for time credits using an exchange platform, which not only brings members utility, but allows them to network as well as develop other problem solving and innovative spin-off projects. The exchange platform is provided by the base organization which has founded the network with idea that by networking people will themselves solve not only their own problems, but also the problems of a region.

P2P Self-Managed

A system which is built by people to transform their unused skills, knowledge, assets, resources and time to offers and requests. Members share their offers and requests for services in order to receive services useful for them. The transactions are managed by members themselves with the help of an exchange tool. This system proceeds from members' perceived need to solve a certain problem within their region.

P2P Virtual

A system which is built by people to transform their unused skills, knowledge, assets, resources and time into offers and requests. Members can share their offers and requests for services in order to obtain services useful for them. The system is managed by members themselves with the help of a computer system. This system proceeds from members' perceived need to solve a certain problem within their region.

P2O

A system aimed at solving particular societal problems within a region by activating people and giving them a tool to resolve the issues themselves. This system motivates members by providing rewards along with a framework to transform members' unused skills, knowledge, assets, resources and time into services exchangeable for time credits to insure that exchanges are fair and equal for everyone. P2O is founded by an organization which has an interest in solving the problems.

P2O Homogenous

A system aimed at solving one particular societal problem within a region by activating people and giving them a tool to resolve the issue themselves. This system motivates members by providing rewards along with a framework to transform members' unused skills, knowledge, assets, resources and time into services exchangeable for time credits to insure that exchanges are fair and equal for everyone. P2O is founded by an organization which has an interest in solving the problems.

P2O Pension Scheme

A system aimed at offering people a way to transform their unused skills, knowledge, assets, resources and time into services in order to help to solve particular regional societal problems by providing equal exchange of time credits as well as the possibility to store them, withdraw them and use them later to cover their needs. This scheme is founded by a trustworthy organization which facilitates the exchanges.

O2O

A system which gives organizations – whether a business, NGO or governmental organizations – an opportunity to transform unused skills, knowledge, assets, resources and time which would otherwise would not be utilized and which is impossible to sell/buy at the moment. These benefits are transformed into offers and requests, and thus can be exchanged for time credits that can be used for what members need to keep their operations efficient, to allow further development and to provide a networking opportunity to facilitate further problem solving, innovations and spin-off projects. O2O is founded by one of organizations concerned and involves the entire regional environment.

O2OP

A system which involves all persons in a region, both physical and legal, and gives them an opportunity to transform their unused skills, knowledge, assets, resources and time into offers and requests which would otherwise would not be utilized and which is impossible to sell/buy at the moment. These benefits are transformed into offers and requests, and thus can be exchanged for time credits that can be used for what members need to keep their operations efficient, to allow further development and to provide a networking opportunity to facilitate further problem solving, innovations and spin-off projects. O2OP is founded by one of organizations concerned and involves the entire regional environment.

TB Networks

A system which allows the equal exchange of offers, requests and resources among any interested Time Banks and is founded and organized by the bodies concerned for the further development of Time Banking.

THE UNIVERSAL MODEL

In previous sections of the book, all elements have been introduced. All nine models categorized in the TB Taxonomy have been analyzed. After a comparison of these full lists of inputs, entities and processes with all the models a universal model was synthetized. This universal model contains all the important parts of the previous models, but brings several innovations. Universal, in this context, means that it contains all necessary shared basic elements, not all encompassing ultimate model merged from all existing models. The models described above are based on real life situations and

Figure 9. The Universal Time Bank model
Source: Author's research.

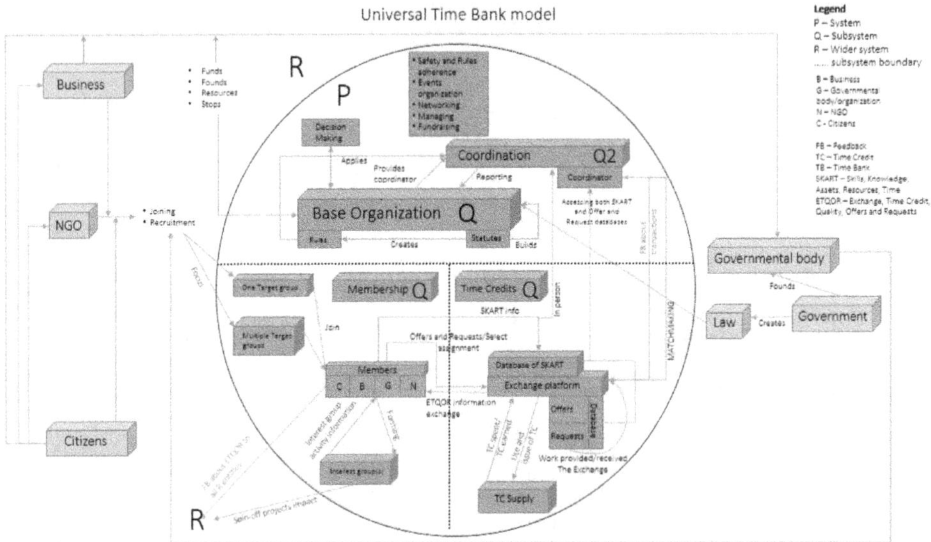

experiences from implementation in various contexts. Usually a TB is founded by a group of enthusiasts without any previous research; in some situations they use a guide from an internet source, a manual of some kind, or advice from those more experienced in the field or from an supportive organization (Valek, 2016). This lack of uniformity is actually one of reasons there has been no complete systemic description of TB, and this lack of one standard description has led to a situation in which TBs are constructed on an almost random basis before founders learn basic facts, such as that they need a coordinator, almost certainly some software program, as well as the fact that people will "somehow" exchange services. This brings us to the point which, although TBs often work (and often fail), the findings of the analysis above show that TBs are systemically in disarray. A synthesis was conducted, again with the use of rich picture design, with results captured in the following chart followed by a description. The decryption clarifies why these elements were made part of the synthesis.

Inputs

The inputs included are based on universal inputs common for all of the TB models. All TBs need to be founded by some entity. They also need funding,

Box 18. The Universal Time Bank model

Universal model of TB						
Inputs	**FB kind**	**P: The system (Transformation process)**				
Funding		**Q: Sub-Systems**				
Founding		**Membership**	**Base Organization**		**Time Credits**	
Joining			**Coordination (is Q within base org)**			
Recruitment		**Entities**	**Entities**	**Coordination Q2**	**Entities**	
Unused SKART		One TG	Statutes	**Entities**	Time Credit supply	
SKART		Multiple TG	Rules	Coordinator	Exchange platform (IT solution)	
Law		Interest groups			Database of SKART	
Physical resources		Members (people and organizations)			Database of Offers and Requests	
		Processes	**Processes**	**Processes**	**Processes**	**FB kind**
		Focus on TG(s)	Coordinates (provides, finds or appoints coordinator)	SKART Info receiving	Earning of TC	+
		Form Interest groups		Accessing the SKART database input/read	Spending of TC	+
		SKART info provide		Accessing the Offer/Req. database input/read	Issue of TC	
		Multiple TG SKART Info provide	Builds rules and statute upon a law	**Matchmaking**	Use of TC	
	+ or -	Feedback about (ETQOR)		SKART INFO use for matchmaking	Work provided/received	
		Provide or Receive Offer/Request Info	Makes strategical decisions	Offer/Request info use for matchmaking	SKART INFO receiving	
				Recruitment	FB about transactions to coordination	
				Networking and Events organizing	Information about ETQOR	
		Select assignment		Fundraising		
		FB about quality to and from members (references)		Managing		
				Security keeping		
		Info about earned/spent TC		Reporting to Base organization		
		R: Wider system entities (Ownership level)		**Processes**		
		Citizens		Founds, funds, resources, joins, can also stop TB existence if founding entity		
		NGO		Founds, Funds, resources, joins, can also stop TB existence if founding entity		
		Businesses		Founds, Funds, resources, joins, can also stop TB existence if founding entity		
	+ or -	Government	Sub-Sub-System	Founds, Funds, resources, provides law frame, joins, can also stop TB existence if founding entity		
			Governmental org.	Founds, Funds, resources, joins, can also stop TB existence if founding entity		
			Law	Provides law framework		

Source: Author's research.

not necessarily significant, although financial backing is recommended to ensure sustainability. They need main inputs in the form of people who may join themselves or be recruited by a coordinator. Their "fuel" is always the unused free SKART, and even ordinary used marketable SKART can be traded (as explained previously). They need a law framework to operate in and to formulate statutes and rules; this ensures both continuity and security. And finally, other physical resources may be provided, e.g. in-kind contributions, which are in essence quite similar to the funding input.

System and Its Sub-Systems

The sub-systems were almost the same in all the TB approaches aside from couple of models in which the base organization was replaced by an exchange platform. Therefore, all four main sub-systems were included in the universal model: Membership, Base Organization, Coordination and Time Credits. The Time Credits themselves in particular were re-structured with a leaner design, i.e. one type used in all situations, including IT Solutions.

MEMBERSHIP

Entities

In all models there was the possibility of focusing on one or multiple target groups, which is natural in that it may start with only one, but when the influence of the TB grows it may attract multiple target groups, or it may start with multiple TBs from the beginning.

Interest groups are one of key structures which might appear in a TB. These bring higher value and actually are usually to be encouraged rather than suppressed. Interest groups usually appear by the activity of members as a result of the co-production upon which TB concept is based.

Members in the universal model may be any body, physical or legal. Citizens, NGOs, businesses and governmental organizations can become members and thus input their SKART to be transformed into offers and requests.

Processes

Focus on target groups and forming interest groups has been explained in earlier sections. In addition to the fact that members also provide their SKART info to the coordinator, and they can register it to the exchange platform as well. This of course, goes for all target groups and all members.

Members provide feedback outside to wider system about ETQOR. They can also access the exchange platform and provide information about their own offers and requests, or they can obtain information about the same from other members. All members select work accordingly. In essence, they have access to ETQOR information from the platform and can also update it. This also includes feedback about quality in the form of references about exchanges on both sides to and from the exchange platform, as well as information about Earned/Spent TC.

Base Organization

This aspect remains unchanged in most models and in the universal one it is the same. The main entities are Statutes and Rules. This relates also to the process by which statutes are created based on local law, and on this basis rules are created for the TB operation.

The base organization also appoints the coordinator and makes strategic decisions.

Coordination

This entity emerges generally as a part of the base organization. In most senses this also represents a full approach by which the main and only entity is the coordinator, who performs key processes for a TB. These processes are:

Receiving SKART information from members if they decide not to record it into the exchange platform themselves and leave it to be attended to by the coordinator.

The coordinator can input or read information about members' SKART and to do the same with the database of offers and requests.

He/she uses this information to perform efficient matchmaking among members who might potentially exchange based on already provided offers and requests, and/ or on SKART information.

Aside from all of the above, the coordinator manages a TB in the sense of doing the administrative work, keeps it safe by actively watching out for and avoiding possible crises, and ensures adherence to TB rules. The coordinator also takes care of any necessary fundraising as well as organizes various events, whether networking (now under the events organization process), education, public relations, etc. Besides this, he/she can use all the knowledge described above to support the recruitment of new members.

The last activity the coordinator performs is to report back to the base organization about the performance of TB, quality of exchanges, etc.

Time Credits

The main difference in structure of this sub-system is the premise that any TB nowadays will use an IT Solution for its operation. This premise will be clarified further in the discussion chapter and is closely related to Information Management Frameworks theory. Therefore, the sub-sub-system of record medium was omitted and now the whole exchange platform is considered to be an IT Solution, i.e. it contains a database of SKART and database of offers and requests as an integral part. Therefore, in the end, this TC sub-system has only two entities: the afore-mentioned exchange platform and TC supply. TC supply has to remain separate, as it is the reference point about amount of TC in circulation.

The processes involved are again earning and spending of TC, both of which, as before, bring positive feedback about the activity of exchanges in TB (reflected in ETQOR). Issuing and use of TC takes place as in previous situations, i.e. issue and use can occur separately on an exchange done. Work provided and received is referred to the database of offers and requests. Information about SKART is inputted to the SKART database by members or a coordinator. Finally, there are two sources of feedback: one tracing transactions to coordination, i.e. the coordinator, who is aware how the TB is performing and can also report it back to the base organization; and the feedback about quality (references of members about exchanges) back to members.

"Bottom Up" Approach

One primary feature considered in relation to the nine original models and to the research experience is the fact that the universal model should be always built from "bottom up." In this way, the point of view – the weltanschauung – proceeds from the side of a Customer in the SSM CATWOE framework (Checkland, 1993). It is very important to note that the synthesis process starts from this side and the universal model is built from the bottom. Experience from the past indicates that a TB which is "installed" from top has considerably less chance of success, as it might (intentionally or not) prioritize the point of view of the implementing institution rather than the real needs of a region. Therefore, it is necessary to involve only the entities and processes that are really needed, not the ones authorities think might be needed.

The identified elements are the basic building blocks of any TB, and by using them and the related processes, it is possible to build a skeleton of a working TB. This skeleton, derived from existing diverse and often unorganized approaches, brings a new point of view. It is clear of any redundancies and offers the possibility of adaptation on the solid ground of a systemic approach. Nevertheless, to underline specific differences its features should be compared with the knowledge gained about contemporary approaches mentioned in the Taxonomy.

REFERENCES

Anthes, G. (2016). Open Source Software No Longer Optional. *Communications of the ACM*, *59*(8), 15–17. doi:10.1145/2949684

Cahn, E. S. (2000). *No more throw-away people: the co-production imperative*. Washington, DC: Essential.

Cahn, E. S., & Rowe, J. (1998). *Time dollars: the new currency that enables Americans to turn their hidden resource--time--into personal security & community renewal*. Chicago: Family Resource Coalition of America.

Checkland, P. (1993). *Systems Thinking, Systems Practice*. Chichester, UK: John Wiley & Sons Ltd.

Checkland, P. (1999). Soft systems methodology: A thirty year retrospective. *Systems Research and Behavioral Science*, *17*(S1), 11–S58. doi:10.1002/1099-1743(200011)17:1+<::AID-SRES374>3.0.CO;2-O

Clement, N., Holbrook, A., Forster, D., Macneil, J., Smith, M., Lyons, K., & Donald, E. M. (2017). Timebanking, Co-production and Normative Principles: Putting Normative Principles into Practice. *International Journal of Community Currency Research, 21*, 37–52.

Fitzgerald, B. (2011). Adopting Open Source Software: A Practical Guide. Cambridge, MA: The MIT Press. doi:10.7551/mitpress/9780262516358.001.0001

Hildreth, P., Kimble, C., & Wright, P. (2000). Communities of practice in the distributed international environment. *Journal of Knowledge Management, 4*(1), 27–38. doi:10.1108/13673270010315920

hourworld.org. (2017). *Hourmobile mobile application on Google Play*. Available: https://play.google.com/store/apps/details?id=edu.psu.ist.mtb_hourworld&hl=en

Lehaney, B. (2004). Beyond Knowledge Management. Idea Group Publishing.

Lesser, E. L., & Storck, J. (2001). Communities of practice and organizational performance. *IBM Systems Journal*, 40.

Lietaer, B. (2001). *The Future of Money*. London: Century, Random House Group Limited.

Linderman, M., Siegel, B., Ouellet, D., Brichacek, J., Haines, S., Chase, G., & O´May, J. (2005). *A Reference Model for Information Management to Support Coalition Information Sharing Needs*. Academic Press.

Maes, R. (1999). *Reconsidering Information Mangagement Through A Generic Framework*. Amsterdam: Universiteit van Amsterdam.

Maes, R., Rijsenbrij, D., Truijens, O., & Goedvolk, H. (2000). *Redefining business - IT alingment through a unified frameworks*. Universiteit van Amsterdam.

Nonaka, I. (1991). *The Knowledge-Creating Company*. Harward Business Review.

Otcenaskova, T., Bures, V., & Mikulecka, J. (2012). Theoretical Fundaments of Knowledge Intensity Modelling. *18th International Business Information Management Association conference*, 1232-1243.

Sacks, M. (2015). Competition Between Open Source and Proprietary Software: Strategies for Survival. *Journal of Management Information Systems*, *32*(3), 268–295. doi:10.1080/07421222.2015.1099391

Tucnik, P., Valek, L., Blecha, P. & Bures, V. (2016). Use of Time Banking as a non-monetary component in agent-based computational economics models. *WSEAS Transactions on Business and Economics,* 13.

Valek, L. (2016). Open Ways for Time Banking Research: Project Management and Beyond. *International Journal of Human Capital and Information Technology Professionals*, *7*(1), 35–47. doi:10.4018/IJHCITP.2016010103

Valek, L. (2018). Three New Directions for Time Banking Research: Information Management, Knowledge Management, and the Open Source Model. In V. Ahuja & S. Rathore (Eds.), *Multidisciplinary Perspectives on Human Capital and Information Technology Professionals*. Hershey, PA: IGI Global. doi:10.4018/978-1-5225-5297-0.ch017

Valek, L. (2015). Time Banks and Knowledge Sharing: Link to the Knowledge Management. *5th International Conference Lumen 2014, Transdisciplinary and Communicative Action (Lumen-Tca 2014),* 809-813.

Valek, L., Kolerova, K., & Otcenaskova, T. (2014). Time banks and clusters: Similarities of sharing framework. *Global Journal on Technology, 6*, 31-36.

Widodo, R. N. S., Lim, H., & Atiquzzaman, M. (2017). SDM: Smart deduplication for mobile cloud storage. *Future Generation Computer Systems-the International Journal of Escience*, *70*, 64–73. doi:10.1016/j.future.2016.06.023

Chapter 5
Usage of TB in Practice

ABSTRACT

This chapter compares the universal TB model with the previous TB interpretation discussed in the first half of the book. Special attention is devoted to 19 selected TB properties ranging from adaptability to universality. Furthermore, tangible results follow in form of potential implications to business administration. Namely, information and knowledge management and information technologies are considered. In the area of information management, a generic time bank information management framework is outlined. In the scope of knowledge management, the function of TB is compared with the phenomenon known as communities of practice. Finally, co-productive behavior of the TB related to open source communities are depicted in the area of information technologies.

SYSTEM CHARACTERISTICS: THE COMPARISON

This sub-chapter is aimed at confronting the system characteristics of previous TB approaches with the new universal approach. By previous approaches, the original nine types described in taxonomy chapter are meant. The characteristics are listed below and for each one there is a short description.

DOI: 10.4018/978-1-5225-6974-9.ch005

Accessibility

Previous Approach

This TB concept is quite accessible in its essence, which proceeds from the basic idea of being a tool for people and organizations to facilitate co-production. The information about how system performs can be usually obtained two ways: either by personal experience witnessing a TB in operation and adopting a pre-existing concept, an approach which can be hardly called conceptual or systemic, or by consulting an idea for a TB with a professional in the area who can provide training and advise. The latter is a much more effective approach, but unfortunately such professionals are present only in countries where the TB concept has been strongly developed.

Universal Approach

As discovered in the breakdown of existing systems, the new system is very simple and can be included in any learning material. This would provide explicit knowledge anywhere in very vivid way, thus making it accessible and understandable even without the presence of an expert.

Adaptability

Previous Approach

Adaptation is a must within a TB. It must be able to adapt to local conditions in order to work. This adaptation may be tricky if the only information available is about one of the aforementioned approaches and if the target situation is very specific.

Universal Approach

As simple as possible, containing only the main essential elements, and can thus be adapted easily.

Affordability

This aspect is the same in both situations. Any of the TB approaches needs a coordinator. It is recommended that it be a paid position to attain sustainability, but it may also be a voluntary position, or the coordinator can earn TC for her / his work. In this sense, the approaches are similar.

Autonomy

Previous Approach

For example, P2O approaches are absolutely bound to their base organizations and cannot be autonomous and TBs are founded by an organization to fulfil its societal role. In general, any TB cannot reliably exist without a base organization and is related to its environment by various ties.

Universal Approach

It is proposed to always have a legal form, thus it is always based on an organization. Nevertheless, this organization is personification of a legal body of a TB. In other words, the organization is the TB, and, as it is stand-alone organization, it is nevertheless created from the needs of the environment and as such the TB will always be dependent on many influences.

Credibility

Previous Approach

Various models of TB have various level of credibility. In general, credibility should be assured by an adherence to rules and laws and by the presence of a coordinator as a security keeper, but there are also approaches where this element is missing.

Universal Approach

Being built on a sound methodology involving knowledge management theory as well as the implementation of all security measures from known approaches, the credibility of new approach should be considerably higher.

Customizability

Previous Approach

Original approaches are clearly already strongly customized, some to the extent that the original ideas of the TB are becoming slightly distorted. This is not necessarily negative, as adaptability is one of main valuable features of TB concept, but it might hinder efforts of potential newcomers to implement the idea successfully.

Universal Approach

Because it is an absolutely fundamental approach built only from basic building blocks of a TB, it allows a clean start without any redundant burden processes or entities which might not be useful. And it allows full customizability for particular local conditions.

Complexity

Previous Approach

Even though TB is in its essence as simple a soft system as may be, the countless modifications have made it hard to grasp from a simple point of view. Also, the rhetoric around TB is confusing, sometimes presenting it as an economic alternative, sometimes a social system, sometimes as a business mutual assistance tool, etc.

Universal Approach

The new approach breaks down TB to its roots and introduces it as it is, i.e. as universal tool for activating and using dormant SKART which is very simple and has less than twenty entities and less than forty processes. Its relation to the complexity rationalization of Time Credit sub-system should be also mentioned. The sub-system was re-organized in order to simplify the needs for an IT Solution and to make system more robust.

Efficiency

Previous Approach

It is very hard to measure the efficiency of a TB system, especially when existing ones use various approaches, and the terminology has not been standardized in any way. Comparison is a tricky issue when each of the existing TBs is slightly different. An evaluation of efficiency may be provided according to certain measures within a TB, but it would hardly be transferable and comparable to others.

Universal Approach

As a "tabula rasa," all processes are clear and can be followed and evaluated easily. After adaptations, this advantage may be lost, but at any rate it might be possible to track adaptations to their core, when the core is known.

Emergence

Is a key feature of TB in general. By putting together people and resources which are facilitated through co-production, it produces multiple outputs and also various spin-off systems.

Equality

Equality is a core value of a TB and this remains unchanged.

Evolvability

Previous Approach

Most of current approaches have already evolved from a simple P2P model. Evolvability goes hand to hand with adaptability and new features can be always added.

Universal Approach

One advantage of this analysis is that it has been undertaken more than three decades after idea of TB was invented. Since then, all the aforementioned approaches have evolved, a situation which facilitates a synthesis of much richer and concentrated ideas than was possible with original P2P TB formulations. In simplification to this basic level, this approach allows a new, more systems based evolution by adaptations.

Inclusiveness

Previous Approach

Some of the current TB approaches are limited to certain target groups or activities, although one of core values is inclusivity, as we are all assets and everyone has something to offer. Thus with the focus on certain target groups, which is natural, all approaches remain very inclusive. If not, TB can transform to some higher, more inclusive level.

Universal Approach

In its universal native form, it allows any member to participate, but this might change with adaptations to local conditions. In other words, TB will be one of most inclusive systems of its kind, but sometimes it must be focused on a specific group of people or organizations, so this does not change compared to original approach.

Interoperability

Previous Approach

Often some IT solutions now offer the possibility to connect various TBs and in doing so the opportunity to network them (as introduced) so they can enrich their sphere of operation. This was a huge step, providing P2O models with completely new possibilities.

Universal Approach

It has an advantage in standardized roots, thus each TB founded by this approach would have the same basic template of entities and processes, a situation which makes it easier to tailor it to any of solution.

Modularity

This is also a key feature of the TB concept in general. As visible from diagrams and charts, it is not a problem to add a new module to the framework of adaptability. This goes universally for original and new approaches.

Predictability

Previous Approach

As a soft system, TB is generally hard to predict. It is based on people (and people as the main element of organizations), which means that system may be quite unpredictable. All nine original approaches are basically defined only by core values and no other obligatory rules are stated.

Universal Approach

By adding a framework with obligatory entities and by defining them, the TB system becomes more predictable, at least in its simplest form. It is still a soft system, but the framework gives it boundaries of operation.

Sustainability

Previous Approach

In the previous nine models some models are characterized by a very low potential toward sustainability, especially those which do not include a coordinator. Also, clear rules should be set, as the existence of a TB is based on trust in its transparency. Experience shows that the lifetime of a TB is often not very long if it is implemented under the wrong conditions, or its basic rules of operation are not adhered to. It then quickly loses trust, and once that starts to happen the end of a TB is usually near.

Universal Approach

By its simplicity and clear rules, the new universal approach should be more robust and would not allow any behavior within the system which would lead to a loss of trust or to any other dangerous developments for a TB.

Tailorability

Previous Approach

In general, it is possible to pick an approach and tailor it to a particular situation, and later adapt the details. This is how it usually works, but without a general universal approach, there might be redundancies which would burden TB operation and might cause disturbances, as outlined above.

Universal Approach

The universal approach is designed to be tailored and adapted.

Transparency

Previous Approach

Although core values and the basic system of operation and exchange are by their nature transparent, execution does not necessarily follow this philosophy. This again results in losing trust, which might lead to an untimely end for a TB.

Universal Approach

Because it is clearly defined, it is also transparent.

Universality

Previous Approach

Each of the adaptations has its own field of operation and only the O2OP model may be called near to universal, as it allows the participation of all wider system entities, but still carries redundancies from its origins.

Universal Approach

Universality is the main feature of the new universal approach, as its aim was to de-construct original approaches and find a universal approach, so it is as universal as it may be, utilizing years of development of the TB idea and taking the best from its various modifications.

IMPLICATIONS TO BUSINESS ADMINISTRATION

In this sub-chapter the results of the research in the context of business administration is outlined. There is a myriad of perspectives that can be applied in this field. Information and knowledge management together with related technologies are exemplified here. Other fields such as human resources management or marketing activities could be easily discussed as well.

Information Management

As noted above, in comparison with all original models the Time Credit sub-system has been simplified in anticipation of a more complex solution, a goal which would cover the needs of TB as a system from point of view of information management (IM). This is not only desirable, but even necessary so that the entities and stakeholders involved have access to information to make optimal decisions (Linderman et al., 2005). As there is no available theory or applicable research which would consider the TB environment, it is necessary to use an approach from the business sphere (Valek, 2018). This is not a serious issue, as now the universal approach to TB is very generic in its nature, so a generic solution to managing the information from both within and outside of a TB system is only natural. The structure comes from the Generic Information Management Framework proposed by (Maes, 1999).

For this book, the structure of the proposed Generic Information Management Framework (GIMF) is ideal, as it considers all important elements. Moreover, the particular cells of the framework are equally important (Maes, 1999). As noted above, the term business might be confusing, so it can be replaced with the word Time Bank for these purposes (Valek, 2018).

To describe the structure of the GIMF, that the columns group (from left to right), business expertise, interpretation of information, and providing technology can be used. The horizontal lines represent from the bottom, the

Figure 1. A generic information management framework
Source: (Maes et al., 2000).

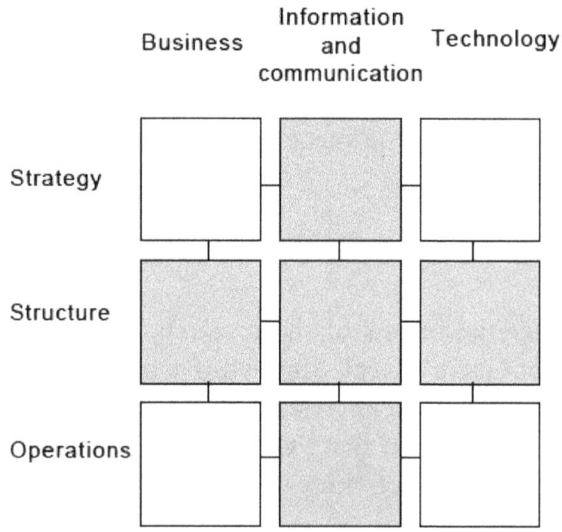

operative, tactical, and strategical levels of these groups (Maes et al., 2000). Based on the analyses and results above, the adaptation of GIMF to the TB environment is as shown in Figure 2.

Time Bank Column

Following the bottom up structure, the TB originates from a need of a target group, which turns to its member base, which in turn provides inputs, most importantly their SKART. In order to root their activity, they must adhere to local law and regulations which would lead to the foundation of a base organization, the Time Bank. When a TB is in operation, members would begin exchanges and they also can form interest groups (Valek, 2018).

Information and Communications Column

The coordinator is the central junction of all information flows, i.e. an information bridge as well as a crossroads. Records about all activities and SKART pass through coordinator so he/she can facilitate the function of the TB. The information is used to provide information to the Base organization for strategic decision making (Valek, 2018).

Figure 2. Generic Time Bank Information Management Framework (GTBIMF)
Source: (Valek, 2018).

	Time Bank	Information and Communication	Technology
Strategy	Law and Regulations – Statutes – Rules Strategic decisions by base organization (organization/community on which a TB is based)	Reporting to base organization	Choosing appropriate IT solution, designing it and/or adapting it to current needs of a TB Both desktop and mobile applications are desirable
Structure	Members Target Groups Inputs including skills, knowledge, assets, resources and time	Coordinator is a center of information flows and connects all elements	Appropriate simple structure which would cover all necessary elements for TB operation
Operations	Exchange activity and feedbacking Forming of interest groups	Proper records of exchanges, time credits, quality and offers and requests, same as, skills, knowledge, assets, resources and time	Main features should be simplicity, ease to use, robustness, accessibility, nearly autopoietic with almost no need for external resources, with running costs very low or free

Technology Column

According to the bottom up approach, members require a simple, inexpensive, robust, accessible and easy to use technological solution. The solution proposed in the universal model is one which would contain all entities of the exchange medium which integrates databases of SKART and Offers and Requests and is related to a TC supply entity to ensure transparency. The final solution depends on availability, but in general it can be designed from scratch. An existing solution can be implemented, if available, or it can be adapted. The selected solution should not only encompass a desktop version, but also contain a mobile application option.

The table introduced above can be called a Generic Time Bank Information Management Framework (GTBIMF) and it is the first of its kind. It not only explains simplifications in Time Credit entity structure, but also proposes a new way of looking at information flows in a TB (Valek, 2018).

Knowledge Management

Probably the strongest relation to knowledge management which was confirmed is knowledge sharing, introduced above as Communities of Practice (CoP) as in

(Valek, 2015), because it is strongly related to the co-production (Cahn, 2000, Cahn and Rowe, 1998, Clement et al., 2017) idea. We can even say it is almost the same, only with the difference that CoPs have in fact been introduced in the literature in a business environment. As knowledge management is mainly about managing how humans can share their knowledge effectively using technical tools where appropriate (Lehaney, 2004), this approach can be applied to any sphere of human activity. In addition, the above-proposed GTBIMF provides a clear framework, not only for the information, but also knowledge sharing. Both CoPs and interest groups in a TB can only exist and operate successfully when members trust each other, and trust is main building block of a TB; trust is secured by rules and by the coordinator's presence. Knowledge is a bond between the social and professional links between people interested in particular areas that enable them to share experience and the understanding of a problem. Membership and choice of community needs to be voluntary, otherwise members may not participate in the knowledge sharing (Lehaney, 2004). If we consider the fact that all TBs are composed of people, whether individuals or people bound to organizations, this description categorically describes how a TB operates in the sense of knowledge sharing. The benefits of CoPs are based mostly on the creation of social capital, which allows the further development of an organization. By being part of CoP, people obtain the benefits of connections and relationships framed in a common context, i.e. the social capital of the community, by which they then have then power to enhance performance of the organization CoP is operating in (Lesser and Storck, 2001). In other words, by giving people within an organization the opportunity to participate in a TB we allow the creation of CoPs or interest groups in this context, thus enhancing performance of the organization. These findings have already been shown in the methodology chapter, but now they are demonstrated by the existence and utter similarity of CoPs and interest groups within TBs (Valek, 2018).

By exchanging services and knowledge in TBs, experiential learning, which relates to tacit knowledge, occurs. Managing this knowledge entails surfacing it (making it explicit in some way) (Nonaka, 1991). Experiential knowledge can also be converted into fluid knowledge so that others can share it (Lehaney, 2004). The knowledge intensity (Otcenaskova et al., 2012) of all participants within knowledge sharing systems, such as TB, is one of the essential keystones of variability of choice within the system. In other words, the more participants of the system know, the more they have to share, thus the more possibilities there are for sharing and, in case of TB, the more opportunities to spend the earned TC they have (Valek, 2015). This idea of Knowledge intensity actually supports the assumption that the more SKART is inputted, the more possibilities TB offers and the more outputs it brings, as knowledge is part of the "sKart" (Valek, 2018).

Another similarity is that face to face meetings and communications are considered to be most important for developing relationships and learning within CoPs (Lesser and Storck, 2001). So it is in Time Banks, in which the personal meeting is even more important than the exchange itself (Valek et al., 2014). That does not mean virtualization is not possible, but again it brings both benefits and risks in both situations. Virtualization or non-personal communication raises a question about the legitimacy of exchanges and brings more obstacles to participation of members (Hildreth et al., 2000), as demonstrated in the analysis.

Based on the paragraphs above, we can say that co-production is a phenomenon which is present both among individuals and in organizations of any kind. We can also say that co-production leads to the creation of CoPs/interest groups, and these provide organizations or a TB with a much greater potential than pure operation based on original "design" of an organization. To clarify the point, TB is not only a complementary economic system, but it contributes a great deal of extra added value in knowledge sharing and co-production, a fact which allows the multiplication of the activities of its members, thus strengthening existing positive feedbacks of a TB system upon the outside wider system environment, and creating completely new solutions and feedbacks with them, e.g. the various kinds of knowledge at the various nodes in a TB

A final remark about information management in a TB is related to the fact that all TB is an actual source of information about exchanges, most of which would never happen without a TB. This kind of "bookkeeping" is a source of information about the transactions in a transparent way. This is due to the fact that even though TC are not money, they are kind of currency, and currency is also an information system (Lietaer, 2001). Practically, this fact means that by implementing TB, authorities can gain an overview about newly uncovered market possibilities, exchanges between people and organizations, needs of society and possible new ideas for innovations (Valek, 2018).

Information Technologies

As mentioned in the introduction, at the present time solutions are mostly limited to adaptations of content management applications, as they offer the required attributes of low to no cost and are generally accessible even to less skilled developers.

Most promising for TB software development seems to be the open source software domain, as it does not require high investments, as a proprietary solution might (Sacks, 2015). And although an original solution needs to be adapted and "forged" into software which would fit needs of a TB, only a minimal investment might be required (Anthes, 2016). On the other hand, this adaptation may also be instituted by co-production. Open source attracts various problem-solving communities and thus forms CoPs on the basis of co-production. To further develop this thought of co-production in an open source environment, we can say that it has been present in the software development area for long time, it is just not called this way. As open source software has opened completely new ways of innovations and markets, the TB concept, working on same principles of co-production, opens new opportunities on a societal level by offering so-called diffusion of innovation (Fitzgerald, 2011). This diffusion offers the possibility of innovation to be communicated among members of a social system, such as a TB or an open source community (Valek, 2018).

Based on open source and with simplicity as one of main attributes, more extensive solutions like ERP systems are "out of the game" already. Larger, more complex systems would be a burden to a TB system, as it needs only a simple virtual "marketplace" and two databases. Plus, the software would be ideally maintenance free and very robust.

On the other hand, a mobile application extension would be welcomed. The hOurworld organization has already been developed for their Time & Talents application called hOurmobile (hOurworld.org, 2017). With the development of mobile applications, it would be advisable to implement a cloud solution to a TB software package to handle the synchronicity of data between units within the system (Widodo et al., 2017).

Another possibility is use of simulation software. If a TB is founded according to the universal approach developed in this book, it is not complicated to follow all flows, gather data and use this data for further simulations, or to take a system as it is and try to simulate it in various contexts without preceding data. An idea has already been proposed to try agent-based simulations to determine the viability of the implementation of TB under various conditions and in various geographical areas (Tucnik et al., 2016). Such simulations can be beneficial, not only to determine the feasibility of a TB set up, but also, if feasibility is demonstrated, to convince stakeholders to support the idea. Simulations can be also done ex-post, to determine possible flaws in operation and avoid them in the future (Valek, 2018).

REFERENCES

Anthes, G. (2016). Open Source Software No Longer Optional. *Communications of the ACM*, *59*(8), 15–17. doi:10.1145/2949684

Cahn, E. S. (2000). *No more throw-away people: the co-production imperative.* Washington, DC: Essential.

Cahn, E. S., & Rowe, J. (1998). *Time dollars: the new currency that enables Americans to turn their hidden resource--time--into personal security & community renewal.* Chicago: Family Resource Coalition of America.

Clement, N., Holbrook, A., Forster, D., Macneil, J., Smith, M., Lyons, K., & Donald, E. M. (2017). Timebanking, Co-production and Normative Principles: Putting Normative Principles into Practice. *International Journal of Community Currency Research, 21*, 37–52.

Fitzgerald, B. (2011). Adopting Open Source Software: A Practical Guide. Cambridge, MA: The MIT Press. doi:10.7551/mitpress/9780262516358.001.0001

Hildreth, P., Kimble, C., & Wright, P. (2000). Communities of practice in the distributed international environment. *Journal of Knowledge Management, 4*(1), 27–38. doi:10.1108/13673270010315920

hourworld.org. (2017). *Hourmobile mobile application on Google Play*. Available: https://play.google.com/store/apps/details?id=edu.psu.ist.mtb_hourworld&hl=en

Lehaney, B. (2004). Beyond Knowledge Managemen. Idea Group Publishing.

Lesser, E. L., & Storck, J. (2001). Communities of practice and organizational performance. *IBM Systems Journal*, 40.

Lietaer, B. (2001). *The Future of Money*. London: Century, Random House Group Limited.

Linderman, M., Siegel, B., Ouellet, D., Brichacek, J., Haines, S., Chase, G., & O´May, J. (2005). *A Reference Model for Information Management to Support Coalition Information Sharing Needs*. Academic Press.

Maes, R. (1999). *Reconsidering Information Mangagement Through A Generic Framework*. Amsterdam: Universiteit van Amsterdam.

Maes, R., Rijsenbrij, D., Truijens, O., & Goedvolk, H. (2000). *Redefining business - IT alingment through a unified frameworks*. Universiteit van Amsterdam.

Nonaka, I. (1991). *The Knowledge-Creating Company*. Harvard Business Review.

Otcenaskova, T., Bures, V., & Mikulecka, J. (2012). Theoretical Fundaments of Knowledge Intensity Modelling. *18th International Business Information Management Association conference*, 1232-1243.

Sacks, M. (2015). Competition Between Open Source and Proprietary Software: Strategies for Survival. *Journal of Management Information Systems, 32*(3), 268–295. doi:10.1080/07421222.2015.1099391

Tucnik, P., Valek, L., Blecha, P., & Bures, V. (2016). Use of Time Banking as a non-monetary component in agent-based computational economics models. *WSEAS Transactions on Business and Economics,* 13.

Valek, L. (2018). Three New Directions for Time Banking Research: Information Management, Knowledge Management, and the Open Source Model. In V. Ahuja & S. Rathore (Eds.), *Multidisciplinary Perspectives on Human Capital and Information Technology Professionals*. Hershey, PA: IGI Global. doi:10.4018/978-1-5225-5297-0.ch017

Valek, L. (2015). Time Banks and Knowledge Sharing: Link to the Knowledge Management. *5th International Conference Lumen 2014, Transdisciplinary and Communicative Action (Lumen-Tca 2014),* 809-813.

Valek, L., Kolerova, K., & Otcenaskova, T. (2014). Time banks and clusters: Similarities of sharing framework. *Global Journal on Technology, 6*, 31-36.

Widodo, R. N. S., Lim, H., & Atiquzzaman, M. (2017). SDM: Smart deduplication for mobile cloud storage. *Future Generation Computer Systems-the International Journal of Escience, 70*, 64–73. doi:10.1016/j.future.2016.06.023

Conclusion

SEARCH QUESTION AND AIM AND OBJECTIVES IN CONCLUSION

Research question: Is there a general approach available for the establishment of Time Banking concept/system?

Aim: Aim of the book is to describe TB as a soft system, identify its elements, analyze them by using variety of methods, which would be able to deal with complexity of the situation, in order to find a universal solution to TB concept which may be used later through most of areas of human activities.

Objectives:

1. To provide a relevant holistic background about TB which would allow the work to progress further to systemic analysis
2. To find an appropriate methodology to analyze TB as a soft system
3. To identify all the elements of TB system
4. To design a universal TB concept based on its core elements
5. To identify the main characteristics of that universal system
6. To identify the main knowledge types and flows in TB related to its core elements
7. To identify the main information flows in TB related to its core elements
8. To determine a context in the information technologies area related to findings

A bottom up process will be applied in this conclusion of aim and objectives and finally the research question will be answered.

Objectives

1. As a complex research problem without much of previous research available, it was not easy to find a context for TB in any science in general. Some sources claim it to be economic system, some that it is purely social, some see it as a complement to socio-economic reality, and some even as an alternative. The relation to many sciences was explored during the entire course of the research, only a fraction of which was mentioned in this book, in order to provide a sufficient background to progress to systemic sciences and systems engineering. A TB was finally identified as a system, and as a soft system, a finding which in the end allowed for further analyses and synthesis, with the final definition presented below.
2. To find a methodology which can be used to define system components of a soft system was another task. In the book, a way to determine the final approach is briefly described, namely Soft System Methodology. This proved to be ideal method to deal with such an unstructured problem as TB turned out to be. Nevertheless, even though SSM is a universal tool for dealing with soft systems, after many iterations of application aside from this book, it was decided to simplify it by taking into account Peter Checkland's (the creator of SSM) reflection, and experiences of others with rich pictures. This led to a completely new design of the rich picture/systemic scheme, which may already include a view of CATWOE, making all approaches much simpler and easy to use. Without any previous background in the categorization of TB approaches (everything found in literature was quite incomplete and burdened by the particular point of view of a researcher), it was necessary to compile knowledge about kinds of TBs, cross-reference this with my own tacit knowledge gained by years of research and, finally, to produce my own taxonomy of TB models. The taxonomy uncovered four main pure models of TB with five more sub-variants, making nine models in total. These nine models were then drafted in a diagram, with system elements isolated and structured into charts.
3. Based on diagrams and charts of original nine models of TB, all inputs into system, all system entities and processes, and outputs were presented and described in detail.
4. After comparing all of these systemic elements, a universal model was drafted and proposed. This universal model combines the main building blocks of elements contained in other approaches, omits exotic and

redundant ones as well as adds the point of view of information and knowledge management, all in order to create a lean and light structure which can be used as a basis for any TB in under any conditions.

5. The main characteristics in the form of systems attributes were described in comparison to the original approaches and the universal approach turned to be superior to them. Nevertheless, it has to be taken in account that the present model was built on the experience of the original approaches, all of which have been tested by trial and error for more than the past three decades.
6. The main flows of information and the main kinds of knowledge which appear in the universal system were identified. In addition, a new generic information framework was proposed, one based on a generic framework for businesses, which makes the TB information management framework first of its kind.
7. The main relationships to information technologies were underlined, especially the relationship of the main core drive force of TB, co-production, and its close relationship to open source software, which actually shows how systemic problem solving is already present in society, not only in the socio-economic context, but also in the information technology domain.

By fulfilling the aforementioned objectives, the aim was reached. A universal approach to TB was isolated from the existing approaches which is now possible to use in any area where TB is intended to be founded as a systemic basis upon which to build local and regional societal problem solving structures. The universal approach is one which is accessible, adaptable, able to evolve and possible to tailor to various situations. It is also affordable, has a certain level of autonomy, is credible due to its transparency and security as well as customizable and modular because of its uncomplicated design. Its nature as a soft system makes it emergent in sense that its outputs surpass outputs of individual systems elements; in addition, it creates an environment for new systems to be born. It possesses low complexity and is efficient, equal and inclusive; due to its standard base it encompasses propositions for interoperability and predictability. And finally, with proper adherence to rules it is sustainable.

The answer to research question is therefore: After researching background and all possible approaches to Time Banking and finding an appropriate methodology for systems analysis a set of basic core components was isolated out of which a universally applicable Time Banking system can be constructed.

Time Bank Definition

Based on the results of this book a general TB definition can be stated:

Time Bank is a socio-economic soft system which gathers people, both individuals and people as part of organizations, to become members of the system, in order to facilitate use of skills, knowledge, assets, resources and time which they are actually not using, or cannot use (especially in a labor market). These skills, knowledge, assets, resources, and time are the main input into the Time Bank as a system where they are transformed into offers and requests which can be utilized by members who require them. This transformation is a source of various outputs, the main one being utility and societal problem-solving in the sense of forming communities which actively seek problems to solve.

From a more general point of view of systems approach the definition would be:

Time Bank is an open concrete continuous and stochastic soft system which tends toward autopoiesis, and which transforms inputs of wider system environment by leverage of knowledge, skills and resources of its sub-systems to multiple outputs with positive feedbacks put back into the wider system environment.

Research Limitations

As is probably clear from the whole text of this book, there were many obstacles and limitations to the research. A main one, the lack of sources and research about TB, was also one of driving motives to start the research in the first place. This book is the very first attempt at all to grasp scientifically the TB concept in such a broad scale and organize it with use of various theories and methods into a strictly defined framework. Most of the available sources were not usable for this book, therefore most of knowledge about operation of TBs and whole problematic situation had to be obtained from the tacit knowledge of various stakeholders around the world. The available information about TB had to be merged with the knowledge obtained as well as had to be related to available theories from other sciences to create a reasonable framework for a TB as a science. This, of course, places a geographical obstacle into the research, as for a more extensive study it would be necessary to travel everywhere a TB operates. In the end, the main stakeholders from the primary involved countries were contacted and interviewed, which brought enough of overview of the situation.

All quantitative research turned to be unserviceable, as mentioned earlier in the short note about statistical analysis, because of the lack of standardization and completely different approaches to data gathering and information management in various TBs. Thus, it was possible to obtain certain information about the operation of a single TBs, but this data was usually incomparable with other ones.

MAIN TANGIBLE RESULTS OF THIS BOOK

Researching Theoretical Background of TB

Aside from the classic books of Edgar S. Cahn and few other books focused on TB as a complementary economic system, few publications (manuals, pamphlets, books, etc.) are available about TB as a model for those who would like to create a comprehensive study about the impact of TB. No research at all can be found about some topics. The TB concept is not directly related to other sciences and the introduction presented by this book is a first, as yet incomplete, step. It was not aim of this book to create a comprehensive study about all the possible sciences TB can be connected to, an undertaking which would require far more information than even another full book could provide. As a by-product, however, this summary can serve as a theoretical starting point for further research in the field.

Summarizing a TB Approach in Systemic Way

Placing TB in the context of systems thinking and systems sciences is also very important for the theoretical base of TB in general. Now that there is a definition of a TB as a soft system which had not been there before, TB can be finally thought of in another way than listing vague descriptions around the five core values.

Creating a TB Taxonomy

Even though some categories of TBs exist, there had been no comprehensive Taxonomy. The taxonomy was built partially on existing categories and partially on the author's own research. It shows main approaches to TB with its many variants and it is based on a systemic approach, which is also unique in the sphere of TB research.

Using a Systemic Methodology to Analyze TB Structure Based on the Taxonomy

To analyze such a vaguely defined soft system as TB requires a special tool. That tool was found in Soft System Methodology and allowed the use of the Taxonomy as a compilation of TB models which may be broken down and analyzed. By that another original output has been produced; an analysis of TB approaches which is based on a proven methodology. Until now all analyses have been based on pure description of a TB without any clear definition or a methodology. This book has brought a new point of view of systems thinking, which facilitated the study TB as a system

Describing Elements of TB, Isolating Which Are Universal and Synthetizing Universal TB System

Based on the above step, it was possible to isolate system elements which are universal for any TB approach and out of these build a universal TB system which may be described, defined, and also implemented under any conditions. The universal approach has also another important feature: because it represents a unified template, it allows further research to be based on systemic grounds and would also facilitate quantitative research as well as computer simulations in the future.

Relation of TB to Business Administration

Knowledge is one of main elements shared in a TB. The findings in relation to knowledge management and even to information technologies have demonstrated that these aspects share common ground, especially in non-formal problem solving areas, in which interest groups of TB and Communities of Practice from the business sphere are practically the same. And both of these are closely related to the community-based phenomena of open source. This relationship has also never been suggested and described before.

Now that a universal TB system has been defined, it is now also possible to relate the concept on an even deeper level to theories of information management. Information flows in TB have never been built around any specific framework, as this aspect has never been researched in depth. One of results of this book is the actual application of the Generic Time Bank Information Management Framework which is to be adhered to when a TB concept is implemented.

Simplifying SSM for Common Use by PQR

Lastly, the SSM approach, as much as it is a universal tool, can be at the same time confusing, as its universal approach presents the opportunity to look at a system from different points of view, its chief aim. In surveying the countless iterations of the application during the research supporting this book, a simplified approach was found which considers one point of view, the perspective of CATWOE. This by-product can be also used for further benefits, especially in the practical application sphere for small businesses in terms of soft system analysis.

FURTHER RESEARCH POSSIBLE

This book was a first step which seeks to open up new possibilities for research. These may take the form of investigations using the humanistic sciences which were touched upon at the beginning of the book. As with any of the related areas, these can be researched in greater depth. TB may be studied from the side of psychology, sociology, anthropology, political sciences, and of course, economics, social work and many other fields. In the more "hard" sciences the door is open for computer simulations, including agent-based simulations of TB in various environments, statistical analyses, mathematical models, software and mobile application design, etc.

And in the systemic sciences further development can proceed, for example, in the direction of systems archetypes, which can be identified by researching currently operational TBs and solutions can be found to solve problems caused by the disturbances these archetypes cause.

Related Readings

To continue our tradition of advancing information science and technology research, we have compiled a list of recommended IGI Global readings. These references will provide additional information and guidance to further enrich your knowledge and assist you with your own research and future publications.

Abtahi, M. S., Behboudi, L., & Hasanabad, H. M. (2017). Factors Affecting Internet Advertising Adoption in Ad Agencies. *International Journal of Innovation in the Digital Economy*, *8*(4), 18–29. doi:10.4018/IJIDE.2017100102

Agrawal, S. (2017). The Impact of Emerging Technologies and Social Media on Different Business(es): Marketing and Management. In O. Rishi & A. Sharma (Eds.), *Maximizing Business Performance and Efficiency Through Intelligent Systems* (pp. 37–49). Hershey, PA: IGI Global. doi:10.4018/978-1-5225-2234-8.ch002

Alnoukari, M., Razouk, R., & Hanano, A. (2016). BSC-SI: A Framework for Integrating Strategic Intelligence in Corporate Strategic Management. *International Journal of Social and Organizational Dynamics in IT*, *5*(2), 1–14. doi:10.4018/IJSODIT.2016070101

Alnoukari, M., Razouk, R., & Hanano, A. (2016). BSC-SI, A Framework for Integrating Strategic Intelligence in Corporate Strategic Management. *International Journal of Strategic Information Technology and Applications*, *7*(1), 32–44. doi:10.4018/IJSITA.2016010103

Altındağ, E. (2016). Current Approaches in Change Management. In A. Goksoy (Ed.), *Organizational Change Management Strategies in Modern Business* (pp. 24–51). Hershey, PA: IGI Global. doi:10.4018/978-1-4666-9533-7.ch002

Alvarez-Dionisi, L. E., Turner, R., & Mittra, M. (2016). Global Project Management Trends. *International Journal of Information Technology Project Management*, *7*(3), 54–73. doi:10.4018/IJITPM.2016070104

Anantharaman, R. N., Rajeswari, K. S., Angusamy, A., & Kuppusamy, J. (2017). Role of Self-Efficacy and Collective Efficacy as Moderators of Occupational Stress Among Software Development Professionals. *International Journal of Human Capital and Information Technology Professionals*, *8*(2), 45–58. doi:10.4018/IJHCITP.2017040103

Aninze, F., El-Gohary, H., & Hussain, J. (2018). The Role of Microfinance to Empower Women: The Case of Developing Countries. *International Journal of Customer Relationship Marketing and Management*, *9*(1), 54–78. doi:10.4018/IJCRMM.2018010104

Arsenijević, O. M., Orčić, D., & Kastratović, E. (2017). Development of an Optimization Tool for Intangibles in SMEs: A Case Study from Serbia with a Pilot Research in the Prestige by Milka Company. In M. Vemić (Ed.), *Optimal Management Strategies in Small and Medium Enterprises* (pp. 320–347). Hershey, PA: IGI Global. doi:10.4018/978-1-5225-1949-2.ch015

Aryanto, V. D., Wismantoro, Y., & Widyatmoko, K. (2018). Implementing Eco-Innovation by Utilizing the Internet to Enhance Firm's Marketing Performance: Study of Green Batik Small and Medium Enterprises in Indonesia. *International Journal of E-Business Research*, *14*(1), 21–36. doi:10.4018/IJEBR.2018010102

Atiku, S. O., & Fields, Z. (2017). Multicultural Orientations for 21st Century Global Leadership. In N. Baporikar (Ed.), *Management Education for Global Leadership* (pp. 28–51). Hershey, PA: IGI Global. doi:10.4018/978-1-5225-1013-0.ch002

Atiku, S. O., & Fields, Z. (2018). Organisational Learning Dimensions and Talent Retention Strategies for the Service Industries. In N. Baporikar (Ed.), *Global Practices in Knowledge Management for Societal and Organizational Development* (pp. 358–381). Hershey, PA: IGI Global. doi:10.4018/978-1-5225-3009-1.ch017

Ávila, L., & Teixeira, L. (2018). The Main Concepts Behind the Dematerialization of Business Processes. In M. Khosrow-Pour, D.B.A. (Ed.), Encyclopedia of Information Science and Technology, Fourth Edition (pp. 888-898). Hershey, PA: IGI Global. doi:10.4018/978-1-5225-2255-3.ch076

Bartens, Y., Chunpir, H. I., Schulte, F., & Voß, S. (2017). Business/IT Alignment in Two-Sided Markets: A COBIT 5 Analysis for Media Streaming Business Models. In S. De Haes & W. Van Grembergen (Eds.), *Strategic IT Governance and Alignment in Business Settings* (pp. 82–111). Hershey, PA: IGI Global. doi:10.4018/978-1-5225-0861-8.ch004

Bashayreh, A. M. (2018). Organizational Culture and Organizational Performance. In W. Lee & F. Sabetzadeh (Eds.), *Contemporary Knowledge and Systems Science* (pp. 50–69). Hershey, PA: IGI Global. doi:10.4018/978-1-5225-5655-8.ch003

Bedford, D. A. (2018). Sustainable Knowledge Management Strategies: Aligning Business Capabilities and Knowledge Management Goals. In N. Baporikar (Ed.), *Global Practices in Knowledge Management for Societal and Organizational Development* (pp. 46–73). Hershey, PA: IGI Global. doi:10.4018/978-1-5225-3009-1.ch003

Benmoussa, F., Nakara, W. A., & Jaouen, A. (2016). The Use of Social Media by SMEs in the Tourism Industry. In I. Lee (Ed.), *Encyclopedia of E-Commerce Development, Implementation, and Management* (pp. 2159–2170). Hershey, PA: IGI Global. doi:10.4018/978-1-4666-9787-4.ch155

Berger, R. (2016). Indigenous Management and Bottom of Pyramid Countries: The Role of National Institutions. In U. Aung & P. Ordoñez de Pablos (Eds.), *Managerial Strategies and Practice in the Asian Business Sector* (pp. 107–123). Hershey, PA: IGI Global. doi:10.4018/978-1-4666-9758-4.ch007

Bharwani, S., & Musunuri, D. (2018). Reflection as a Process From Theory to Practice. In M. Khosrow-Pour, D.B.A. (Ed.), Encyclopedia of Information Science and Technology, Fourth Edition (pp. 1529-1539). Hershey, PA: IGI Global. doi:10.4018/978-1-5225-2255-3.ch132

Bhatt, G. D., Wang, Z., & Rodger, J. A. (2017). Information Systems Capabilities and Their Effects on Competitive Advantages: A Study of Chinese Companies. *Information Resources Management Journal, 30*(3), 41–57. doi:10.4018/IRMJ.2017070103

Bhushan, M., & Yadav, A. (2017). Concept of Cloud Computing in ESB. In R. Bhadoria, N. Chaudhari, G. Tomar, & S. Singh (Eds.), *Exploring Enterprise Service Bus in the Service-Oriented Architecture Paradigm* (pp. 116–127). Hershey, PA: IGI Global. doi:10.4018/978-1-5225-2157-0.ch008

Bhushan, S. (2017). System Dynamics Base-Model of Humanitarian Supply Chain (HSCM) in Disaster Prone Eco-Communities of India: A Discussion on Simulation and Scenario Results. *International Journal of System Dynamics Applications, 6*(3), 20–37. doi:10.4018/IJSDA.2017070102

Biswas, A., & De, A. K. (2017). On Development of a Fuzzy Stochastic Programming Model with Its Application to Business Management. In S. Trivedi, S. Dey, A. Kumar, & T. Panda (Eds.), *Handbook of Research on Advanced Data Mining Techniques and Applications for Business Intelligence* (pp. 353–378). Hershey, PA: IGI Global. doi:10.4018/978-1-5225-2031-3.ch021

Bücker, J., & Ernste, K. (2018). Use of Brand Heroes in Strategic Reputation Management: The Case of Bacardi, Adidas, and Daimler. In A. Erdemir (Ed.), *Reputation Management Techniques in Public Relations* (pp. 126–150). Hershey, PA: IGI Global. doi:10.4018/978-1-5225-3619-2.ch007

Bureš, V. (2018). Industry 4.0 From the Systems Engineering Perspective: Alternative Holistic Framework Development. In R. Brunet-Thornton & F. Martinez (Eds.), *Analyzing the Impacts of Industry 4.0 in Modern Business Environments* (pp. 199–223). Hershey, PA: IGI Global. doi:10.4018/978-1-5225-3468-6.ch011

Buzady, Z. (2017). Resolving the Magic Cube of Effective Case Teaching: Benchmarking Case Teaching Practices in Emerging Markets – Insights from the Central European University Business School, Hungary. In D. Latusek (Ed.), *Case Studies as a Teaching Tool in Management Education* (pp. 79–103). Hershey, PA: IGI Global. doi:10.4018/978-1-5225-0770-3.ch005

Campatelli, G., Richter, A., & Stocker, A. (2016). Participative Knowledge Management to Empower Manufacturing Workers. *International Journal of Knowledge Management, 12*(4), 37–50. doi:10.4018/IJKM.2016100103

Căpusneanu, S., & Topor, D. I. (2018). Business Ethics and Cost Management in SMEs: Theories of Business Ethics and Cost Management Ethos. In I. Oncioiu (Ed.), *Ethics and Decision-Making for Sustainable Business Practices* (pp. 109–127). Hershey, PA: IGI Global. doi:10.4018/978-1-5225-3773-1.ch007

Carneiro, A. (2016). Maturity in Health Organization Information Systems: Metrics and Privacy Perspectives. *International Journal of Privacy and Health Information Management, 4*(2), 1–18. doi:10.4018/IJPHIM.2016070101

Chan, R. L., Mo, P. L., & Moon, K. K. (2018). Strategic and Tactical Measures in Managing Enterprise Risks: A Study of the Textile and Apparel Industry. In K. Strang, M. Korstanje, & N. Vajjhala (Eds.), *Research, Practices, and Innovations in Global Risk and Contingency Management* (pp. 1–19). Hershey, PA: IGI Global. doi:10.4018/978-1-5225-4754-9.ch001

Chandan, H. C. (2016). Motivations and Challenges of Female Entrepreneurship in Developed and Developing Economies. In N. Baporikar (Ed.), *Handbook of Research on Entrepreneurship in the Contemporary Knowledge-Based Global Economy* (pp. 260–286). Hershey, PA: IGI Global. doi:10.4018/978-1-4666-8798-1.ch012

Charlier, S. D., Burke-Smalley, L. A., & Fisher, S. L. (2018). Undergraduate Programs in the U.S: A Contextual and Content-Based Analysis. In J. Mendy (Ed.), *Teaching Human Resources and Organizational Behavior at the College Level* (pp. 26–57). Hershey, PA: IGI Global. doi:10.4018/978-1-5225-2820-3.ch002

Chaudhuri, S. (2016). Application of Web-Based Geographical Information System (GIS) in E-Business. In U. Panwar, R. Kumar, & N. Ray (Eds.), *Handbook of Research on Promotional Strategies and Consumer Influence in the Service Sector* (pp. 389–405). Hershey, PA: IGI Global. doi:10.4018/978-1-5225-0143-5.ch023

Choudhuri, P. S. (2016). An Empirical Study on the Quality of Services Offered by the Private Life Insurers in Burdwan. In U. Panwar, R. Kumar, & N. Ray (Eds.), *Handbook of Research on Promotional Strategies and Consumer Influence in the Service Sector* (pp. 31–55). Hershey, PA: IGI Global. doi:10.4018/978-1-5225-0143-5.ch002

Dahlberg, T., Kivijärvi, H., & Saarinen, T. (2017). IT Investment Consistency and Other Factors Influencing the Success of IT Performance. In S. De Haes & W. Van Grembergen (Eds.), *Strategic IT Governance and Alignment in Business Settings* (pp. 176–208). Hershey, PA: IGI Global. doi:10.4018/978-1-5225-0861-8.ch007

Damnjanović, A. M. (2017). Knowledge Management Optimization through IT and E-Business Utilization: A Qualitative Study on Serbian SMEs. In M. Vemić (Ed.), *Optimal Management Strategies in Small and Medium Enterprises* (pp. 249–267). Hershey, PA: IGI Global. doi:10.4018/978-1-5225-1949-2.ch012

Daneshpour, H. (2017). Integrating Sustainable Development into Project Portfolio Management through Application of Open Innovation. In M. Vemić (Ed.), *Optimal Management Strategies in Small and Medium Enterprises* (pp. 370–387). Hershey, PA: IGI Global. doi:10.4018/978-1-5225-1949-2.ch017

Daniel, A. D., & Reis de Castro, V. (2018). Entrepreneurship Education: How to Measure the Impact on Nascent Entrepreneurs. In A. Carrizo Moreira, J. Guilherme Leitão Dantas, & F. Manuel Valente (Eds.), *Nascent Entrepreneurship and Successful New Venture Creation* (pp. 85–110). Hershey, PA: IGI Global. doi:10.4018/978-1-5225-2936-1.ch004

David, F., van der Sijde, P., & van den Besselaar, P. (2016). Enterpreneurial Incentives, Obstacles, and Management in University-Business Co-Operation: The Case of Indonesia. In J. Saiz-Álvarez (Ed.), *Handbook of Research on Social Entrepreneurship and Solidarity Economics* (pp. 499–518). Hershey, PA: IGI Global. doi:10.4018/978-1-5225-0097-1.ch024

David, R., Swami, B. N., & Tangirala, S. (2018). Ethics Impact on Knowledge Management in Organizational Development: A Case Study. In N. Baporikar (Ed.), *Global Practices in Knowledge Management for Societal and Organizational Development* (pp. 19–45). Hershey, PA: IGI Global. doi:10.4018/978-1-5225-3009-1.ch002

Delias, P., & Lakiotaki, K. (2018). Discovering Process Horizontal Boundaries to Facilitate Process Comprehension. *International Journal of Operations Research and Information Systems*, *9*(2), 1–31. doi:10.4018/IJORIS.2018040101

Denholm, J., & Lee-Davies, L. (2018). Success Factors for Games in Business and Project Management. In *Enhancing Education and Training Initiatives Through Serious Games* (pp. 34–68). Hershey, PA: IGI Global. doi:10.4018/978-1-5225-3689-5.ch002

Deshpande, M. (2017). Best Practices in Management Institutions for Global Leadership: Policy Aspects. In N. Baporikar (Ed.), *Management Education for Global Leadership* (pp. 1–27). Hershey, PA: IGI Global. doi:10.4018/978-1-5225-1013-0.ch001

Deshpande, M. (2018). Policy Perspectives for SMEs Knowledge Management. In N. Baporikar (Ed.), *Knowledge Integration Strategies for Entrepreneurship and Sustainability* (pp. 23–46). Hershey, PA: IGI Global. doi:10.4018/978-1-5225-5115-7.ch002

Dezdar, S. (2017). ERP Implementation Projects in Asian Countries: A Comparative Study on Iran and China. *International Journal of Information Technology Project Management, 8*(3), 52–68. doi:10.4018/IJITPM.2017070104

Domingos, D., Martinho, R., & Varajão, J. (2016). Controlled Flexibility in Healthcare Processes: A BPMN-Extension Approach. In M. Cruz-Cunha, I. Miranda, R. Martinho, & R. Rijo (Eds.), *Encyclopedia of E-Health and Telemedicine* (pp. 521–535). Hershey, PA: IGI Global. doi:10.4018/978-1-4666-9978-6.ch040

Domingos, D., Respício, A., & Martinho, R. (2017). Reliability of IoT-Aware BPMN Healthcare Processes. In C. Reis & M. Maximiano (Eds.), *Internet of Things and Advanced Application in Healthcare* (pp. 214–248). Hershey, PA: IGI Global. doi:10.4018/978-1-5225-1820-4.ch008

Dosumu, O., Hussain, J., & El-Gohary, H. (2017). An Exploratory Study of the Impact of Government Policies on the Development of Small and Medium Enterprises in Developing Countries: The Case of Nigeria. *International Journal of Customer Relationship Marketing and Management, 8*(4), 51–62. doi:10.4018/IJCRMM.2017100104

Durst, S., Bruns, G., & Edvardsson, I. R. (2017). Retaining Knowledge in Smaller Building and Construction Firms. *International Journal of Knowledge and Systems Science, 8*(3), 1–12. doi:10.4018/IJKSS.2017070101

Edvardsson, I. R., & Durst, S. (2017). Outsourcing, Knowledge, and Learning: A Critical Review. *International Journal of Knowledge-Based Organizations, 7*(2), 13–26. doi:10.4018/IJKBO.2017040102

Edwards, J. S. (2018). Integrating Knowledge Management and Business Processes. In M. Khosrow-Pour, D.B.A. (Ed.), Encyclopedia of Information Science and Technology, Fourth Edition (pp. 5046-5055). Hershey, PA: IGI Global. doi:10.4018/978-1-5225-2255-3.ch437

Ejiogu, A. O. (2018). Economics of Farm Management. In *Agricultural Finance and Opportunities for Investment and Expansion* (pp. 56–72). Hershey, PA: IGI Global. doi:10.4018/978-1-5225-3059-6.ch003

Ekanem, I., & Abiade, G. E. (2018). Factors Influencing the Use of E-Commerce by Small Enterprises in Nigeria. *International Journal of ICT Research in Africa and the Middle East, 7*(1), 37–53. doi:10.4018/IJICTRAME.2018010103

Ekanem, I., & Alrossais, L. A. (2017). Succession Challenges Facing Family Businesses in Saudi Arabia. In P. Zgheib (Ed.), *Entrepreneurship and Business Innovation in the Middle East* (pp. 122–146). Hershey, PA: IGI Global. doi:10.4018/978-1-5225-2066-5.ch007

El Faquih, L., & Fredj, M. (2017). Ontology-Based Framework for Quality in Configurable Process Models. *Journal of Electronic Commerce in Organizations, 15*(2), 48–60. doi:10.4018/JECO.2017040104

El-Gohary, H., & El-Gohary, Z. (2016). An Attempt to Explore Electronic Marketing Adoption and Implementation Aspects in Developing Countries: The Case of Egypt. *International Journal of Customer Relationship Marketing and Management, 7*(4), 1–26. doi:10.4018/IJCRMM.2016100101

Entico, G. J. (2016). Knowledge Management and the Medical Health Librarians: A Perception Study. In J. Yap, M. Perez, M. Ayson, & G. Entico (Eds.), *Special Library Administration, Standardization and Technological Integration* (pp. 52–77). Hershey, PA: IGI Global. doi:10.4018/978-1-4666-9542-9.ch003

Faisal, M. N., & Talib, F. (2017). Building Ambidextrous Supply Chains in SMEs: How to Tackle the Barriers? *International Journal of Information Systems and Supply Chain Management, 10*(4), 80–100. doi:10.4018/IJISSCM.2017100105

Fernandes, T. M., Gomes, J., & Romão, M. (2017). Investments in E-Government: A Benefit Management Case Study. *International Journal of Electronic Government Research, 13*(3), 1–17. doi:10.4018/IJEGR.2017070101

Fouda, F. A. (2016). A Suggested Curriculum in Career Education to Develop Business Secondary Schools Students' Career Knowledge Management Domains and Professional Thinking. *International Journal of Technology Diffusion*, *7*(2), 42–62. doi:10.4018/IJTD.2016040103

Gallardo-Vázquez, D., & Pajuelo-Moreno, M. L. (2016). How Spanish Universities are Promoting Entrepreneurship through Your Own Lines of Teaching and Research? In L. Carvalho (Ed.), *Handbook of Research on Entrepreneurial Success and its Impact on Regional Development* (pp. 431–454). Hershey, PA: IGI Global. doi:10.4018/978-1-4666-9567-2.ch019

Gao, S. S., Oreal, S., & Zhang, J. (2018). Contemporary Financial Risk Management Perceptions and Practices of Small-Sized Chinese Businesses. In I. Management Association (Ed.), Global Business Expansion: Concepts, Methodologies, Tools, and Applications (pp. 917-931). Hershey, PA: IGI Global. doi:10.4018/978-1-5225-5481-3.ch041

Garg, R., & Berning, S. C. (2017). Indigenous Chinese Management Philosophies: Key Concepts and Relevance for Modern Chinese Firms. In B. Christiansen & G. Koc (Eds.), *Transcontinental Strategies for Industrial Development and Economic Growth* (pp. 43–57). Hershey, PA: IGI Global. doi:10.4018/978-1-5225-2160-0.ch003

Gencer, Y. G. (2017). Supply Chain Management in Retailing Business. In U. Akkucuk (Ed.), *Ethics and Sustainability in Global Supply Chain Management* (pp. 197–210). Hershey, PA: IGI Global. doi:10.4018/978-1-5225-2036-8.ch011

Giacosa, E. (2016). Innovation in Luxury Fashion Businesses as a Means for the Regional Development. In L. Carvalho (Ed.), *Handbook of Research on Entrepreneurial Success and its Impact on Regional Development* (pp. 206–222). Hershey, PA: IGI Global. doi:10.4018/978-1-4666-9567-2.ch010

Giacosa, E. (2018). The Increasing of the Regional Development Thanks to the Luxury Business Innovation. In L. Carvalho (Ed.), *Handbook of Research on Entrepreneurial Ecosystems and Social Dynamics in a Globalized World* (pp. 260–273). Hershey, PA: IGI Global. doi:10.4018/978-1-5225-3525-6.ch011

Gianni, M., & Gotzamani, K. (2016). Integrated Management Systems and Information Management Systems: Common Threads. In P. Papajorgji, F. Pinet, A. Guimarães, & J. Papathanasiou (Eds.), *Automated Enterprise Systems for Maximizing Business Performance* (pp. 195–214). Hershey, PA: IGI Global. doi:10.4018/978-1-4666-8841-4.ch011

Gianni, M., Gotzamani, K., & Linden, I. (2016). How a BI-wise Responsible Integrated Management System May Support Food Traceability. *International Journal of Decision Support System Technology*, *8*(2), 1–17. doi:10.4018/IJDSST.2016040101

Glykas, M., & George, J. (2017). Quality and Process Management Systems in the UAE Maritime Industry. *International Journal of Productivity Management and Assessment Technologies*, *5*(1), 20–39. doi:10.4018/IJPMAT.2017010102

Glykas, M., Valiris, G., Kokkinaki, A., & Koutsoukou, Z. (2018). Banking Business Process Management Implementation. *International Journal of Productivity Management and Assessment Technologies*, *6*(1), 50–69. doi:10.4018/IJPMAT.2018010104

Gomes, J., & Romão, M. (2017). The Balanced Scorecard: Keeping Updated and Aligned with Today's Business Trends. *International Journal of Productivity Management and Assessment Technologies*, *5*(2), 1–15. doi:10.4018/IJPMAT.2017070101

Gomes, J., & Romão, M. (2017). Aligning Information Systems and Technology with Benefit Management and Balanced Scorecard. In S. De Haes & W. Van Grembergen (Eds.), *Strategic IT Governance and Alignment in Business Settings* (pp. 112–131). Hershey, PA: IGI Global. doi:10.4018/978-1-5225-0861-8.ch005

Grefen, P., & Turetken, O. (2017). Advanced Business Process Management in Networked E-Business Scenarios. *International Journal of E-Business Research*, *13*(4), 70–104. doi:10.4018/IJEBR.2017100105

Haider, A., & Saetang, S. (2017). Strategic IT Alignment in Service Sector. In S. Rozenes & Y. Cohen (Eds.), *Handbook of Research on Strategic Alliances and Value Co-Creation in the Service Industry* (pp. 231–258). Hershey, PA: IGI Global. doi:10.4018/978-1-5225-2084-9.ch012

Haider, A., & Tang, S. S. (2016). Maximising Value Through IT and Business Alignment: A Case of IT Governance Institutionalisation at a Thai Bank. *International Journal of Technology Diffusion*, *7*(3), 33–58. doi:10.4018/IJTD.2016070104

Hajilari, A. B., Ghadaksaz, M., & Fasghandis, G. S. (2017). Assessing Organizational Readiness for Implementing ERP System Using Fuzzy Expert System Approach. *International Journal of Enterprise Information Systems*, *13*(1), 67–85. doi:10.4018/IJEIS.2017010105

Haldorai, A., Ramu, A., & Murugan, S. (2018). Social Aware Cognitive Radio Networks: Effectiveness of Social Networks as a Strategic Tool for Organizational Business Management. In H. Bansal, G. Shrivastava, G. Nguyen, & L. Stanciu (Eds.), *Social Network Analytics for Contemporary Business Organizations* (pp. 188–202). Hershey, PA: IGI Global. doi:10.4018/978-1-5225-5097-6.ch010

Hall, O. P. Jr. (2017). Social Media Driven Management Education. *International Journal of Knowledge-Based Organizations*, *7*(2), 43–59. doi:10.4018/IJKBO.2017040104

Hanifah, H., Halim, H. A., Ahmad, N. H., & Vafaei-Zadeh, A. (2017). Innovation Culture as a Mediator Between Specific Human Capital and Innovation Performance Among Bumiputera SMEs in Malaysia. In N. Ahmad, T. Ramayah, H. Halim, & S. Rahman (Eds.), *Handbook of Research on Small and Medium Enterprises in Developing Countries* (pp. 261–279). Hershey, PA: IGI Global. doi:10.4018/978-1-5225-2165-5.ch012

Hartlieb, S., & Silvius, G. (2017). Handling Uncertainty in Project Management and Business Development: Similarities and Differences. In Y. Raydugin (Ed.), *Handbook of Research on Leveraging Risk and Uncertainties for Effective Project Management* (pp. 337–362). Hershey, PA: IGI Global. doi:10.4018/978-1-5225-1790-0.ch016

Hass, K. B. (2017). Living on the Edge: Managing Project Complexity. In Y. Raydugin (Ed.), *Handbook of Research on Leveraging Risk and Uncertainties for Effective Project Management* (pp. 177–201). Hershey, PA: IGI Global. doi:10.4018/978-1-5225-1790-0.ch009

Hassan, A., & Privitera, D. S. (2016). Google AdSense as a Mobile Technology in Education. In J. Holland (Ed.), *Wearable Technology and Mobile Innovations for Next-Generation Education* (pp. 200–223). Hershey, PA: IGI Global. doi:10.4018/978-1-5225-0069-8.ch011

Hassan, A., & Rahimi, R. (2016). Consuming "Innovation" in Tourism: Augmented Reality as an Innovation Tool in Digital Tourism Marketing. In N. Pappas & I. Bregoli (Eds.), *Global Dynamics in Travel, Tourism, and Hospitality* (pp. 130–147). Hershey, PA: IGI Global. doi:10.4018/978-1-5225-0201-2.ch008

Hawking, P., & Carmine Sellitto, C. (2017). Developing an Effective Strategy for Organizational Business Intelligence. In M. Tavana (Ed.), *Enterprise Information Systems and the Digitalization of Business Functions* (pp. 222–237). Hershey, PA: IGI Global. doi:10.4018/978-1-5225-2382-6.ch010

Hawking, P., & Sellitto, C. (2017). A Fast-Moving Consumer Goods Company and Business Intelligence Strategy Development. *International Journal of Enterprise Information Systems*, *13*(2), 22–33. doi:10.4018/IJEIS.2017040102

Hawking, P., & Sellitto, C. (2017). Business Intelligence Strategy: Two Case Studies. *International Journal of Business Intelligence Research*, *8*(2), 17–30. doi:10.4018/IJBIR.2017070102

Haynes, J. D., Arockiasamy, S., Al Rashdi, M., & Al Rashdi, S. (2016). Business and E Business Strategies for Coopetition and Thematic Management as a Sustained Basis for Ethics and Social Responsibility in Emerging Markets. In M. Al-Shammari & H. Masri (Eds.), *Ethical and Social Perspectives on Global Business Interaction in Emerging Markets* (pp. 25–39). Hershey, PA: IGI Global. doi:10.4018/978-1-4666-9864-2.ch002

Hee, W. J., Jalleh, G., Lai, H., & Lin, C. (2017). E-Commerce and IT Projects: Evaluation and Management Issues in Australian and Taiwanese Hospitals. *International Journal of Public Health Management and Ethics*, *2*(1), 69–90. doi:10.4018/IJPHME.2017010104

Hernandez, A. A. (2018). Exploring the Factors to Green IT Adoption of SMEs in the Philippines. *Journal of Cases on Information Technology*, *20*(2), 49–66. doi:10.4018/JCIT.2018040104

Hernandez, A. A., & Ona, S. E. (2016). Green IT Adoption: Lessons from the Philippines Business Process Outsourcing Industry. *International Journal of Social Ecology and Sustainable Development*, *7*(1), 1–34. doi:10.4018/IJSESD.2016010101

Hollman, A., Bickford, S., & Hollman, T. (2017). Cyber InSecurity: A Post-Mortem Attempt to Assess Cyber Problems from IT and Business Management Perspectives. *Journal of Cases on Information Technology*, *19*(3), 42–70. doi:10.4018/JCIT.2017070104

Igbinakhase, I. (2017). Responsible and Sustainable Management Practices in Developing and Developed Business Environments. In Z. Fields (Ed.), *Collective Creativity for Responsible and Sustainable Business Practice* (pp. 180–207). Hershey, PA: IGI Global. doi:10.4018/978-1-5225-1823-5.ch010

Ilahi, L., Ghannouchi, S. A., & Martinho, R. (2016). A Business Process Management Approach to Home Healthcare Processes: On the Gap between Intention and Reality. In M. Cruz-Cunha, I. Miranda, R. Martinho, & R. Rijo (Eds.), *Encyclopedia of E-Health and Telemedicine* (pp. 439–457). Hershey, PA: IGI Global. doi:10.4018/978-1-4666-9978-6.ch035

Iwata, J. J., & Hoskins, R. G. (2017). Managing Indigenous Knowledge in Tanzania: A Business Perspective. In P. Jain & N. Mnjama (Eds.), *Managing Knowledge Resources and Records in Modern Organizations* (pp. 198–214). Hershey, PA: IGI Global. doi:10.4018/978-1-5225-1965-2.ch012

Jabeen, F., Ahmad, S. Z., & Alkaabi, S. (2016). The Internationalization Decision-Making of United Arab Emirates Family Businesses. In N. Zakaria, A. Abdul-Talib, & N. Osman (Eds.), *Handbook of Research on Impacts of International Business and Political Affairs on the Global Economy* (pp. 1–22). Hershey, PA: IGI Global. doi:10.4018/978-1-4666-9806-2.ch001

Jain, P. (2017). Ethical and Legal Issues in Knowledge Management Life-Cycle in Business. In P. Jain & N. Mnjama (Eds.), *Managing Knowledge Resources and Records in Modern Organizations* (pp. 82–101). Hershey, PA: IGI Global. doi:10.4018/978-1-5225-1965-2.ch006

Jamali, D., Abdallah, H., & Matar, F. (2016). Opportunities and Challenges for CSR Mainstreaming in Business Schools. *International Journal of Technology and Educational Marketing*, *6*(2), 1–29. doi:10.4018/IJTEM.2016070101

James, S., & Hauli, E. (2017). Holistic Management Education at Tanzanian Rural Development Planning Institute. In N. Baporikar (Ed.), *Management Education for Global Leadership* (pp. 112–136). Hershey, PA: IGI Global. doi:10.4018/978-1-5225-1013-0.ch006

Janošková, M., Csikósová, A., & Čulková, K. (2018). Measurement of Company Performance as Part of Its Strategic Management. In R. Leon (Ed.), *Managerial Strategies for Business Sustainability During Turbulent Times* (pp. 309–335). Hershey, PA: IGI Global. doi:10.4018/978-1-5225-2716-9.ch017

Jean-Vasile, A., & Alecu, A. (2017). Theoretical and Practical Approaches in Understanding the Influences of Cost-Productivity-Profit Trinomial in Contemporary Enterprises. In A. Jean Vasile & D. Nicolò (Eds.), *Sustainable Entrepreneurship and Investments in the Green Economy* (pp. 28–62). Hershey, PA: IGI Global. doi:10.4018/978-1-5225-2075-7.ch002

Jha, D. G. (2016). Preparing for Information Technology Driven Changes. In S. Tiwari & L. Nafees (Eds.), *Innovative Management Education Pedagogies for Preparing Next-Generation Leaders* (pp. 258–274). Hershey, PA: IGI Global. doi:10.4018/978-1-4666-9691-4.ch015

Joia, L. A., & Correia, J. C. (2018). CIO Competencies From the IT Professional Perspective: Insights From Brazil. *Journal of Global Information Management, 26*(2), 74–103. doi:10.4018/JGIM.2018040104

Juma, A., & Mzera, N. (2017). Knowledge Management and Records Management and Competitive Advantage in Business. In P. Jain & N. Mnjama (Eds.), *Managing Knowledge Resources and Records in Modern Organizations* (pp. 15–28). Hershey, PA: IGI Global. doi:10.4018/978-1-5225-1965-2.ch002

K., I., & A, V. (2018). Monitoring and Auditing in the Cloud. In K. Munir (Ed.), *Cloud Computing Technologies for Green Enterprises* (pp. 318-350). Hershey, PA: IGI Global. doi:10.4018/978-1-5225-3038-1.ch013

Kabra, G., Ghosh, V., & Ramesh, A. (2018). Enterprise Integrated Business Process Management and Business Intelligence Framework for Business Process Sustainability. In A. Paul, D. Bhattacharyya, & S. Anand (Eds.), *Green Initiatives for Business Sustainability and Value Creation* (pp. 228–238). Hershey, PA: IGI Global. doi:10.4018/978-1-5225-2662-9.ch010

Kaoud, M. (2017). Investigation of Customer Knowledge Management: A Case Study Research. *International Journal of Service Science, Management, Engineering, and Technology, 8*(2), 12–22. doi:10.4018/IJSSMET.2017040102

Kara, M. E., & Fırat, S. Ü. (2016). Sustainability, Risk, and Business Intelligence in Supply Chains. In M. Erdoğdu, T. Arun, & I. Ahmad (Eds.), *Handbook of Research on Green Economic Development Initiatives and Strategies* (pp. 501–538). Hershey, PA: IGI Global. doi:10.4018/978-1-5225-0440-5.ch022

Katuu, S. (2018). A Comparative Assessment of Enterprise Content Management Maturity Models. In N. Gwangwava & M. Mutingi (Eds.), *E-Manufacturing and E-Service Strategies in Contemporary Organizations* (pp. 93–118). Hershey, PA: IGI Global. doi:10.4018/978-1-5225-3628-4.ch005

Khan, M. A. (2016). MNEs Management Strategies in Developing Countries: Establishing the Context. In M. Khan (Ed.), *Multinational Enterprise Management Strategies in Developing Countries* (pp. 1–33). Hershey, PA: IGI Global. doi:10.4018/978-1-5225-0276-0.ch001

Khan, M. A. (2016). Operational Approaches in Organizational Structure: A Case for MNEs in Developing Countries. In M. Khan (Ed.), *Multinational Enterprise Management Strategies in Developing Countries* (pp. 129–151). Hershey, PA: IGI Global. doi:10.4018/978-1-5225-0276-0.ch007

Kinnunen, S., Ylä-Kujala, A., Marttonen-Arola, S., Kärri, T., & Baglee, D. (2018). Internet of Things in Asset Management: Insights from Industrial Professionals and Academia. *International Journal of Service Science, Management, Engineering, and Technology*, *9*(2), 104–119. doi:10.4018/IJSSMET.2018040105

Klein, A. Z., Sabino de Freitas, A., Machado, L., Freitas, J. C. Jr, Graziola, P. G. Jr, & Schlemmer, E. (2017). Virtual Worlds Applications for Management Education. In L. Tomei (Ed.), *Exploring the New Era of Technology-Infused Education* (pp. 279–299). Hershey, PA: IGI Global. doi:10.4018/978-1-5225-1709-2.ch017

Kożuch, B., & Jabłoński, A. (2017). Adopting the Concept of Business Models in Public Management. In M. Lewandowski & B. Kożuch (Eds.), *Public Sector Entrepreneurship and the Integration of Innovative Business Models* (pp. 10–46). Hershey, PA: IGI Global. doi:10.4018/978-1-5225-2215-7.ch002

Kumar, J., Adhikary, A., & Jha, A. (2017). Small Active Investors' Perceptions and Preferences Towards Tax Saving Mutual Fund Schemes in Eastern India: An Empirical Note. *International Journal of Asian Business and Information Management*, *8*(2), 35–45. doi:10.4018/IJABIM.2017040103

Lassoued, Y., Bouzguenda, L., & Mahmoud, T. (2016). Context-Aware Business Process Versions Management. *International Journal of e-Collaboration, 12*(3), 7–33. doi:10.4018/IJeC.2016070102

Lavassani, K. M., & Movahedi, B. (2017). Applications Driven Information Systems: Beyond Networks toward Business Ecosystems. *International Journal of Innovation in the Digital Economy, 8*(1), 61–75. doi:10.4018/IJIDE.2017010104

Lazzareschi, V. H., & Brito, M. S. (2017). Strategic Information Management: Proposal of Business Project Model. In G. Jamil, A. Soares, & C. Pessoa (Eds.), *Handbook of Research on Information Management for Effective Logistics and Supply Chains* (pp. 59–88). Hershey, PA: IGI Global. doi:10.4018/978-1-5225-0973-8.ch004

Lederer, M., Kurz, M., & Lazarov, P. (2017). Usage and Suitability of Methods for Strategic Business Process Initiatives: A Multi Case Study Research. *International Journal of Productivity Management and Assessment Technologies, 5*(1), 40–51. doi:10.4018/IJPMAT.2017010103

Lee, I. (2017). A Social Enterprise Business Model and a Case Study of Pacific Community Ventures (PCV). In V. Potocan, M. Üngan, & Z. Nedelko (Eds.), *Handbook of Research on Managerial Solutions in Non-Profit Organizations* (pp. 182–204). Hershey, PA: IGI Global. doi:10.4018/978-1-5225-0731-4.ch009

Lee, L. J., & Leu, J. (2016). Exploring the Effectiveness of IT Application and Value Method in the Innovation Performance of Enterprise. *International Journal of Enterprise Information Systems, 12*(2), 47–65. doi:10.4018/IJEIS.2016040104

Lee, Y. (2016). Alignment Effect of Entrepreneurial Orientation and Marketing Orientation on Firm Performance. *International Journal of Customer Relationship Marketing and Management, 7*(4), 58–69. doi:10.4018/IJCRMM.2016100104

Leon, L. A., Seal, K. C., Przasnyski, Z. H., & Wiedenman, I. (2017). Skills and Competencies Required for Jobs in Business Analytics: A Content Analysis of Job Advertisements Using Text Mining. *International Journal of Business Intelligence Research, 8*(1), 1–25. doi:10.4018/IJBIR.2017010101

Leu, J., Lee, L. J., & Krischke, A. (2016). Value Engineering-Based Method for Implementing the ISO14001 System in the Green Supply Chains. *International Journal of Strategic Decision Sciences, 7*(4), 1–20. doi:10.4018/IJSDS.2016100101

Levy, C. L., & Elias, N. I. (2017). SOHO Users' Perceptions of Reliability and Continuity of Cloud-Based Services. In M. Moore (Ed.), *Cybersecurity Breaches and Issues Surrounding Online Threat Protection* (pp. 248–287). Hershey, PA: IGI Global. doi:10.4018/978-1-5225-1941-6.ch011

Levy, M. (2018). Change Management Serving Knowledge Management and Organizational Development: Reflections and Review. In N. Baporikar (Ed.), *Global Practices in Knowledge Management for Societal and Organizational Development* (pp. 256–270). Hershey, PA: IGI Global. doi:10.4018/978-1-5225-3009-1.ch012

Lewandowski, M. (2017). Public Organizations and Business Model Innovation: The Role of Public Service Design. In M. Lewandowski & B. Kożuch (Eds.), *Public Sector Entrepreneurship and the Integration of Innovative Business Models* (pp. 47–72). Hershey, PA: IGI Global. doi:10.4018/978-1-5225-2215-7.ch003

Lhannaoui, H., Kabbaj, M. I., & Bakkoury, Z. (2017). A Survey of Risk-Aware Business Process Modelling. *International Journal of Risk and Contingency Management, 6*(3), 14–26. doi:10.4018/IJRCM.2017070102

Li, J., Sun, W., Jiang, W., Yang, H., & Zhang, L. (2017). How the Nature of Exogenous Shocks and Crises Impact Company Performance?: The Effects of Industry Characteristics. *International Journal of Risk and Contingency Management, 6*(4), 40–55. doi:10.4018/IJRCM.2017100103

Lu, C., & Liu, S. (2016). Cultural Tourism O2O Business Model Innovation-A Case Study of CTrip. *Journal of Electronic Commerce in Organizations, 14*(2), 16–31. doi:10.4018/JECO.2016040102

Machen, B., Hosseini, M. R., Wood, A., & Bakhshi, J. (2016). An Investigation into using SAP-PS as a Multidimensional Project Control System (MPCS). *International Journal of Enterprise Information Systems, 12*(2), 66–81. doi:10.4018/IJEIS.2016040105

Malega, P. (2017). Small and Medium Enterprises in the Slovak Republic: Status and Competitiveness of SMEs in the Global Markets and Possibilities of Optimization. In M. Vemić (Ed.), *Optimal Management Strategies in Small and Medium Enterprises* (pp. 102–124). Hershey, PA: IGI Global. doi:10.4018/978-1-5225-1949-2.ch006

Malewska, K. M. (2017). Intuition in Decision-Making on the Example of a Non-Profit Organization. In V. Potocan, M. Üngan, & Z. Nedelko (Eds.), *Handbook of Research on Managerial Solutions in Non-Profit Organizations* (pp. 378–399). Hershey, PA: IGI Global. doi:10.4018/978-1-5225-0731-4.ch018

Maroofi, F. (2017). Entrepreneurial Orientation and Organizational Learning Ability Analysis for Innovation and Firm Performance. In N. Baporikar (Ed.), *Innovation and Shifting Perspectives in Management Education* (pp. 144–165). Hershey, PA: IGI Global. doi:10.4018/978-1-5225-1019-2.ch007

Martins, P. V., & Zacarias, M. (2017). A Web-based Tool for Business Process Improvement. *International Journal of Web Portals*, *9*(2), 68–84. doi:10.4018/IJWP.2017070104

Matthies, B., & Coners, A. (2017). Exploring the Conceptual Nature of e-Business Projects. *Journal of Electronic Commerce in Organizations*, *15*(3), 33–63. doi:10.4018/JECO.2017070103

McKee, J. (2018). Architecture as a Tool to Solve Business Planning Problems. In M. Khosrow-Pour, D.B.A. (Ed.), Encyclopedia of Information Science and Technology, Fourth Edition (pp. 573-586). Hershey, PA: IGI Global. doi:10.4018/978-1-5225-2255-3.ch050

McMurray, A. J., Cross, J., & Caponecchia, C. (2018). The Risk Management Profession in Australia: Business Continuity Plan Practices. In N. Bajgoric (Ed.), *Always-On Enterprise Information Systems for Modern Organizations* (pp. 112–129). Hershey, PA: IGI Global. doi:10.4018/978-1-5225-3704-5.ch006

Meddah, I. H., & Belkadi, K. (2018). Mining Patterns Using Business Process Management. In R. Hamou (Ed.), *Handbook of Research on Biomimicry in Information Retrieval and Knowledge Management* (pp. 78–89). Hershey, PA: IGI Global. doi:10.4018/978-1-5225-3004-6.ch005

Mendes, L. (2017). TQM and Knowledge Management: An Integrated Approach Towards Tacit Knowledge Management. In D. Jaziri-Bouagina & G. Jamil (Eds.), *Handbook of Research on Tacit Knowledge Management for Organizational Success* (pp. 236–263). Hershey, PA: IGI Global. doi:10.4018/978-1-5225-2394-9.ch009

Mnjama, N. M. (2017). Preservation of Recorded Information in Public and Private Sector Organizations. In P. Jain & N. Mnjama (Eds.), *Managing Knowledge Resources and Records in Modern Organizations* (pp. 149–167). Hershey, PA: IGI Global. doi:10.4018/978-1-5225-1965-2.ch009

Mokoqama, M., & Fields, Z. (2017). Principles of Responsible Management Education (PRME): Call for Responsible Management Education. In Z. Fields (Ed.), *Collective Creativity for Responsible and Sustainable Business Practice* (pp. 229–241). Hershey, PA: IGI Global. doi:10.4018/978-1-5225-1823-5.ch012

Muniapan, B. (2017). Philosophy and Management: The Relevance of Vedanta in Management. In P. Ordóñez de Pablos (Ed.), *Managerial Strategies and Solutions for Business Success in Asia* (pp. 124–139). Hershey, PA: IGI Global. doi:10.4018/978-1-5225-1886-0.ch007

Muniapan, B., Gregory, M. L., & Ling, L. A. (2016). Marketing Education in Sarawak: Looking at It from the Employers' Viewpoint. In B. Smith & A. Porath (Eds.), *Global Perspectives on Contemporary Marketing Education* (pp. 112–130). Hershey, PA: IGI Global. doi:10.4018/978-1-4666-9784-3.ch008

Murad, S. E., & Dowaji, S. (2017). Using Value-Based Approach for Managing Cloud-Based Services. In A. Turuk, B. Sahoo, & S. Addya (Eds.), *Resource Management and Efficiency in Cloud Computing Environments* (pp. 33–60). Hershey, PA: IGI Global. doi:10.4018/978-1-5225-1721-4.ch002

Mutahar, A. M., Daud, N. M., Thurasamy, R., Isaac, O., & Abdulsalam, R. (2018). The Mediating of Perceived Usefulness and Perceived Ease of Use: The Case of Mobile Banking in Yemen. *International Journal of Technology Diffusion*, *9*(2), 21–40. doi:10.4018/IJTD.2018040102

Naidoo, V. (2017). E-Learning and Management Education at African Universities. In N. Baporikar (Ed.), *Management Education for Global Leadership* (pp. 181–201). Hershey, PA: IGI Global. doi:10.4018/978-1-5225-1013-0.ch009

Naidoo, V., & Igbinakhase, I. (2018). Opportunities and Challenges of Knowledge Retention in SMEs. In N. Baporikar (Ed.), *Knowledge Integration Strategies for Entrepreneurship and Sustainability* (pp. 70–94). Hershey, PA: IGI Global. doi:10.4018/978-1-5225-5115-7.ch004

Nayak, S., & Prabhu, N. (2017). Paradigm Shift in Management Education: Need for a Cross Functional Perspective. In N. Baporikar (Ed.), *Management Education for Global Leadership* (pp. 241–255). Hershey, PA: IGI Global. doi:10.4018/978-1-5225-1013-0.ch012

Ndede-Amadi, A. A. (2016). Student Interest in the IS Specialization as Predictor of the Success Potential of New Information Systems Programmes within the Schools of Business in Kenyan Public Universities. *International Journal of Information Systems and Social Change*, *7*(2), 63–79. doi:10.4018/IJISSC.2016040104

Nedelko, Z., & Potocan, V. (2016). Management Practices for Processes Optimization: Case of Slovenia. In G. Alor-Hernández, C. Sánchez-Ramírez, & J. García-Alcaraz (Eds.), *Handbook of Research on Managerial Strategies for Achieving Optimal Performance in Industrial Processes* (pp. 545–561). Hershey, PA: IGI Global. doi:10.4018/978-1-5225-0130-5.ch025

Nedelko, Z., & Potocan, V. (2017). Management Solutions in Non-Profit Organizations: Case of Slovenia. In V. Potocan, M. Üngan, & Z. Nedelko (Eds.), *Handbook of Research on Managerial Solutions in Non-Profit Organizations* (pp. 1–22). Hershey, PA: IGI Global. doi:10.4018/978-1-5225-0731-4.ch001

Nedelko, Z., & Potocan, V. (2017). Priority of Management Tools Utilization among Managers: International Comparison. In V. Wang (Ed.), *Encyclopedia of Strategic Leadership and Management* (pp. 1083–1094). Hershey, PA: IGI Global. doi:10.4018/978-1-5225-1049-9.ch075

Nedelko, Z., Raudeliūnienė, J., & Črešnar, R. (2018). Knowledge Dynamics in Supply Chain Management. In N. Baporikar (Ed.), *Knowledge Integration Strategies for Entrepreneurship and Sustainability* (pp. 150–166). Hershey, PA: IGI Global. doi:10.4018/978-1-5225-5115-7.ch008

Nguyen, H. T., & Hipsher, S. A. (2018). Innovation and Creativity Used by Private Sector Firms in a Resources-Constrained Environment. In S. Hipsher (Ed.), *Examining the Private Sector's Role in Wealth Creation and Poverty Reduction* (pp. 219–238). Hershey, PA: IGI Global. doi:10.4018/978-1-5225-3117-3.ch010

Nycz, M., & Pólkowski, Z. (2016). Business Intelligence as a Modern IT Supporting Management of Local Government Units in Poland. *International Journal of Knowledge and Systems Science*, *7*(4), 1–18. doi:10.4018/IJKSS.2016100101

Obaji, N. O., Senin, A. A., & Olugu, M. U. (2016). Supportive Government Policy as a Mechanism for Business Incubation Performance in Nigeria. *International Journal of Information Systems and Social Change*, *7*(4), 52–66. doi:10.4018/IJISSC.2016100103

Obicci, P. A. (2017). Risk Sharing in a Partnership. In *Risk Management Strategies in Public-Private Partnerships* (pp. 115–152). Hershey, PA: IGI Global. doi:10.4018/978-1-5225-2503-5.ch004

Obidallah, W. J., & Raahemi, B. (2017). Managing Changes in Service Oriented Virtual Organizations: A Structural and Procedural Framework to Facilitate the Process of Change. *Journal of Electronic Commerce in Organizations*, *15*(1), 59–83. doi:10.4018/JECO.2017010104

Ojasalo, J., & Ojasalo, K. (2016). Service Logic Business Model Canvas for Lean Development of SMEs and Start-Ups. In N. Baporikar (Ed.), *Handbook of Research on Entrepreneurship in the Contemporary Knowledge-Based Global Economy* (pp. 217–243). Hershey, PA: IGI Global. doi:10.4018/978-1-4666-8798-1.ch010

Ojo, O. (2017). Impact of Innovation on the Entrepreneurial Success in Selected Business Enterprises in South-West Nigeria. *International Journal of Innovation in the Digital Economy*, *8*(2), 29–38. doi:10.4018/IJIDE.2017040103

Okdinawati, L., Simatupang, T. M., & Sunitiyoso, Y. (2017). Multi-Agent Reinforcement Learning for Value Co-Creation of Collaborative Transportation Management (CTM). *International Journal of Information Systems and Supply Chain Management*, *10*(3), 84–95. doi:10.4018/IJISSCM.2017070105

Ortner, E., Mevius, M., Wiedmann, P., & Kurz, F. (2016). Design of Interactional Decision Support Applications for E-Participation in Smart Cities. *International Journal of Electronic Government Research*, *12*(2), 18–38. doi:10.4018/IJEGR.2016040102

Pal, K. (2018). Building High Quality Big Data-Based Applications in Supply Chains. In A. Kumar & S. Saurav (Eds.), *Supply Chain Management Strategies and Risk Assessment in Retail Environments* (pp. 1–24). Hershey, PA: IGI Global. doi:10.4018/978-1-5225-3056-5.ch001

Palos-Sanchez, P. R., & Correia, M. B. (2018). Perspectives of the Adoption of Cloud Computing in the Tourism Sector. In J. Rodrigues, C. Ramos, P. Cardoso, & C. Henriques (Eds.), *Handbook of Research on Technological Developments for Cultural Heritage and eTourism Applications* (pp. 377–400). Hershey, PA: IGI Global. doi:10.4018/978-1-5225-2927-9.ch018

Parry, V. K., & Lind, M. L. (2016). Alignment of Business Strategy and Information Technology Considering Information Technology Governance, Project Portfolio Control, and Risk Management. *International Journal of Information Technology Project Management*, *7*(4), 21–37. doi:10.4018/IJITPM.2016100102

Pashkova, N., Trujillo-Barrera, A., Apostolakis, G., Van Dijk, G., Drakos, P. D., & Baourakis, G. (2016). Business Management Models of Microfinance Institutions (MFIs) in Africa: A Study into Their Enabling Environments. *International Journal of Food and Beverage Manufacturing and Business Models*, *1*(2), 63–82. doi:10.4018/IJFBMBM.2016070105

Patiño, B. E. (2017). New Generation Management by Convergence and Individual Identity: A Systemic and Human-Oriented Approach. In N. Baporikar (Ed.), *Innovation and Shifting Perspectives in Management Education* (pp. 119–143). Hershey, PA: IGI Global. doi:10.4018/978-1-5225-1019-2.ch006

Pawliczek, A., & Rössler, M. (2017). Knowledge of Management Tools and Systems in SMEs: Knowledge Transfer in Management. In A. Bencsik (Ed.), *Knowledge Management Initiatives and Strategies in Small and Medium Enterprises* (pp. 180–203). Hershey, PA: IGI Global. doi:10.4018/978-1-5225-1642-2.ch009

Pejic-Bach, M., Omazic, M. A., Aleksic, A., & Zoroja, J. (2018). Knowledge-Based Decision Making: A Multi-Case Analysis. In R. Leon (Ed.), *Managerial Strategies for Business Sustainability During Turbulent Times* (pp. 160–184). Hershey, PA: IGI Global. doi:10.4018/978-1-5225-2716-9.ch009

Perano, M., Hysa, X., & Calabrese, M. (2018). Strategic Planning, Cultural Context, and Business Continuity Management: Business Cases in the City of Shkoder. In A. Presenza & L. Sheehan (Eds.), *Geopolitics and Strategic Management in the Global Economy* (pp. 57–77). Hershey, PA: IGI Global. doi:10.4018/978-1-5225-2673-5.ch004

Pereira, R., Mira da Silva, M., & Lapão, L. V. (2017). IT Governance Maturity Patterns in Portuguese Healthcare. In S. De Haes & W. Van Grembergen (Eds.), *Strategic IT Governance and Alignment in Business Settings* (pp. 24–52). Hershey, PA: IGI Global. doi:10.4018/978-1-5225-0861-8.ch002

Perez-Uribe, R., & Ocampo-Guzman, D. (2016). Conflict within Colombian Family Owned SMEs: An Explosive Blend between Feelings and Business. In J. Saiz-Álvarez (Ed.), *Handbook of Research on Social Entrepreneurship and Solidarity Economics* (pp. 329–354). Hershey, PA: IGI Global. doi:10.4018/978-1-5225-0097-1.ch017

Pérez-Uribe, R. I., Torres, D. A., Jurado, S. P., & Prada, D. M. (2018). Cloud Tools for the Development of Project Management in SMEs. In R. Perez-Uribe, C. Salcedo-Perez, & D. Ocampo-Guzman (Eds.), *Handbook of Research on Intrapreneurship and Organizational Sustainability in SMEs* (pp. 95–120). Hershey, PA: IGI Global. doi:10.4018/978-1-5225-3543-0.ch005

Petrisor, I., & Cozmiuc, D. (2017). Global Supply Chain Management Organization at Siemens in the Advent of Industry 4.0. In L. Saglietto & C. Cezanne (Eds.), *Global Intermediation and Logistics Service Providers* (pp. 123–142). Hershey, PA: IGI Global. doi:10.4018/978-1-5225-2133-4.ch007

Pierce, J. M., Velliaris, D. M., & Edwards, J. (2017). A Living Case Study: A Journey Not a Destination. In N. Silton (Ed.), *Exploring the Benefits of Creativity in Education, Media, and the Arts* (pp. 158–178). Hershey, PA: IGI Global. doi:10.4018/978-1-5225-0504-4.ch008

Radosavljevic, M., & Andjelkovic, A. (2017). Multi-Criteria Decision Making Approach for Choosing Business Process for the Improvement: Upgrading of the Six Sigma Methodology. In J. Stanković, P. Delias, S. Marinković, & S. Rochhia (Eds.), *Tools and Techniques for Economic Decision Analysis* (pp. 225–247). Hershey, PA: IGI Global. doi:10.4018/978-1-5225-0959-2.ch011

Radovic, V. M. (2017). Corporate Sustainability and Responsibility and Disaster Risk Reduction: A Serbian Overview. In M. Camilleri (Ed.), *CSR 2.0 and the New Era of Corporate Citizenship* (pp. 147–164). Hershey, PA: IGI Global. doi:10.4018/978-1-5225-1842-6.ch008

Raghunath, K. M., Devi, S. L., & Patro, C. S. (2018). Impact of Risk Assessment Models on Risk Factors: A Holistic Outlook. In K. Strang, M. Korstanje, & N. Vajjhala (Eds.), *Research, Practices, and Innovations in Global Risk and Contingency Management* (pp. 134–153). Hershey, PA: IGI Global. doi:10.4018/978-1-5225-4754-9.ch008

Raman, A., & Goyal, D. P. (2017). Extending IMPLEMENT Framework for Enterprise Information Systems Implementation to Information System Innovation. In M. Tavana (Ed.), *Enterprise Information Systems and the Digitalization of Business Functions* (pp. 137–177). Hershey, PA: IGI Global. doi:10.4018/978-1-5225-2382-6.ch007

Rao, Y., & Zhang, Y. (2017). The Construction and Development of Academic Library Digital Special Subject Databases. In L. Ruan, Q. Zhu, & Y. Ye (Eds.), *Academic Library Development and Administration in China* (pp. 163–183). Hershey, PA: IGI Global. doi:10.4018/978-1-5225-0550-1.ch010

Ravasan, A. Z., Mohammadi, M. M., & Hamidi, H. (2018). An Investigation Into the Critical Success Factors of Implementing Information Technology Service Management Frameworks. In K. Jakobs (Ed.), *Corporate and Global Standardization Initiatives in Contemporary Society* (pp. 200–218). Hershey, PA: IGI Global. doi:10.4018/978-1-5225-5320-5.ch009

Renna, P., Izzo, C., & Romaniello, T. (2016). The Business Process Management Systems to Support Continuous Improvements. In W. Nuninger & J. Châtelet (Eds.), *Handbook of Research on Quality Assurance and Value Management in Higher Education* (pp. 237–256). Hershey, PA: IGI Global. doi:10.4018/978-1-5225-0024-7.ch009

Rezaie, S., Mirabedini, S. J., & Abtahi, A. (2018). Designing a Model for Implementation of Business Intelligence in the Banking Industry. *International Journal of Enterprise Information Systems, 14*(1), 77–103. doi:10.4018/IJEIS.2018010105

Riccò, R. (2016). Diversity Management: Bringing Equality, Equity, and Inclusion in the Workplace. In J. Prescott (Ed.), *Handbook of Research on Race, Gender, and the Fight for Equality* (pp. 335–359). Hershey, PA: IGI Global. doi:10.4018/978-1-5225-0047-6.ch015

Romano, L., Grimaldi, R., & Colasuonno, F. S. (2017). Demand Management as a Success Factor in Project Portfolio Management. In L. Romano (Ed.), *Project Portfolio Management Strategies for Effective Organizational Operations* (pp. 202–219). Hershey, PA: IGI Global. doi:10.4018/978-1-5225-2151-8.ch008

Rostek, K. B. (2016). Risk Management: Role and Importance in Business Organization. In D. Jakóbczak (Ed.), *Analyzing Risk through Probabilistic Modeling in Operations Research* (pp. 149–178). Hershey, PA: IGI Global. doi:10.4018/978-1-4666-9458-3.ch007

Rouhani, S., & Savoji, S. R. (2016). A Success Assessment Model for BI Tools Implementation: An Empirical Study of Banking Industry. *International Journal of Business Intelligence Research*, *7*(1), 25–44. doi:10.4018/IJBIR.2016010103

Ruan, Z. (2016). A Corpus-Based Functional Analysis of Complex Nominal Groups in Written Business Discourse: The Case of "Business". *International Journal of Computer-Assisted Language Learning and Teaching*, *6*(2), 74–90. doi:10.4018/IJCALLT.2016040105

Ruhi, U. (2018). Towards an Interdisciplinary Socio-Technical Definition of Virtual Communities. In M. Khosrow-Pour, D.B.A. (Ed.), Encyclopedia of Information Science and Technology, Fourth Edition (pp. 4278-4295). Hershey, PA: IGI Global. doi:10.4018/978-1-5225-2255-3.ch371

Ryan, J., Doster, B., Daily, S., & Lewis, C. (2016). A Case Study Perspective for Balanced Perioperative Workflow Achievement through Data-Driven Process Improvement. *International Journal of Healthcare Information Systems and Informatics*, *11*(3), 19–41. doi:10.4018/IJHISI.2016070102

Safari, M. R., & Jiang, Q. (2018). The Theory and Practice of IT Governance Maturity and Strategies Alignment: Evidence From Banking Industry. *Journal of Global Information Management*, *26*(2), 127–146. doi:10.4018/JGIM.2018040106

Sahoo, J., Pati, B., & Mohanty, B. (2017). Knowledge Management as an Academic Discipline: An Assessment. In B. Gunjal (Ed.), *Managing Knowledge and Scholarly Assets in Academic Libraries* (pp. 99–126). Hershey, PA: IGI Global. doi:10.4018/978-1-5225-1741-2.ch005

Saini, D. (2017). Relevance of Teaching Values and Ethics in Management Education. In N. Baporikar (Ed.), *Management Education for Global Leadership* (pp. 90–111). Hershey, PA: IGI Global. doi:10.4018/978-1-5225-1013-0.ch005

Sambhanthan, A. (2017). Assessing and Benchmarking Sustainability in Organisations: An Integrated Conceptual Model. *International Journal of Systems and Service-Oriented Engineering*, *7*(4), 22–43. doi:10.4018/IJSSOE.2017100102

Sambhanthan, A., & Potdar, V. (2017). A Study of the Parameters Impacting Sustainability in Information Technology Organizations. *International Journal of Knowledge-Based Organizations*, *7*(3), 27–39. doi:10.4018/IJKBO.2017070103

Sánchez-Fernández, M. D., & Manríquez, M. R. (2018). The Entrepreneurial Spirit Based on Social Values: The Digital Generation. In P. Isaias & L. Carvalho (Eds.), *User Innovation and the Entrepreneurship Phenomenon in the Digital Economy* (pp. 173–193). Hershey, PA: IGI Global. doi:10.4018/978-1-5225-2826-5.ch009

Sanchez-Ruiz, L., & Blanco, B. (2017). Process Management for SMEs: Barriers, Enablers, and Benefits. In M. Vemić (Ed.), *Optimal Management Strategies in Small and Medium Enterprises* (pp. 293–319). Hershey, PA: IGI Global. doi:10.4018/978-1-5225-1949-2.ch014

Sanz, L. F., Gómez-Pérez, J., & Castillo-Martinez, A. (2018). Analysis of the European ICT Competence Frameworks. In V. Ahuja & S. Rathore (Eds.), *Multidisciplinary Perspectives on Human Capital and Information Technology Professionals* (pp. 225–245). Hershey, PA: IGI Global. doi:10.4018/978-1-5225-5297-0.ch012

Sarvepalli, A., & Godin, J. (2017). Business Process Management in the Classroom. *Journal of Cases on Information Technology*, *19*(2), 17–28. doi:10.4018/JCIT.2017040102

Satpathy, B., & Muniapan, B. (2016). Ancient Wisdom for Transformational Leadership and Its Insights from the Bhagavad-Gita. In U. Aung & P. Ordoñez de Pablos (Eds.), *Managerial Strategies and Practice in the Asian Business Sector* (pp. 1–10). Hershey, PA: IGI Global. doi:10.4018/978-1-4666-9758-4.ch001

Saygili, E. E., Ozturkoglu, Y., & Kocakulah, M. C. (2017). End Users' Perceptions of Critical Success Factors in ERP Applications. *International Journal of Enterprise Information Systems, 13*(4), 58–75. doi:10.4018/IJEIS.2017100104

Saygili, E. E., & Saygili, A. T. (2017). Contemporary Issues in Enterprise Information Systems: A Critical Review of CSFs in ERP Implementations. In M. Tavana (Ed.), *Enterprise Information Systems and the Digitalization of Business Functions* (pp. 120–136). Hershey, PA: IGI Global. doi:10.4018/978-1-5225-2382-6.ch006

Seidenstricker, S., & Antonino, A. (2018). Business Model Innovation-Oriented Technology Management for Emergent Technologies. In M. Khosrow-Pour, D.B.A. (Ed.), Encyclopedia of Information Science and Technology, Fourth Edition (pp. 4560-4569). Hershey, PA: IGI Global. doi:10.4018/978-1-5225-2255-3.ch396

Senaratne, S., & Gunarathne, A. D. (2017). Excellence Perspective for Management Education from a Global Accountants' Hub in Asia. In N. Baporikar (Ed.), *Management Education for Global Leadership* (pp. 158–180). Hershey, PA: IGI Global. doi:10.4018/978-1-5225-1013-0.ch008

Sensuse, D. I., & Cahyaningsih, E. (2018). Knowledge Management Models: A Summative Review. *International Journal of Information Systems in the Service Sector, 10*(1), 71–100. doi:10.4018/IJISSS.2018010105

Sensuse, D. I., Wibowo, W. C., & Cahyaningsih, E. (2016). Indonesian Government Knowledge Management Model: A Theoretical Model. *Information Resources Management Journal, 29*(1), 91–108. doi:10.4018/irmj.2016010106

Seth, M., Goyal, D., & Kiran, R. (2017). Diminution of Impediments in Implementation of Supply Chain Management Information System for Enhancing its Effectiveness in Indian Automobile Industry. *Journal of Global Information Management, 25*(3), 1–20. doi:10.4018/JGIM.2017070101

Seyal, A. H., & Rahman, M. N. (2017). Investigating Impact of Inter-Organizational Factors in Measuring ERP Systems Success: Bruneian Perspectives. In M. Tavana (Ed.), *Enterprise Information Systems and the Digitalization of Business Functions* (pp. 178–204). Hershey, PA: IGI Global. doi:10.4018/978-1-5225-2382-6.ch008

Shaikh, A. A., & Karjaluoto, H. (2016). On Some Misconceptions Concerning Digital Banking and Alternative Delivery Channels. *International Journal of E-Business Research, 12*(3), 1–16. doi:10.4018/IJEBR.2016070101

Shams, S. M. (2016). Stakeholder Relationship Management in Online Business and Competitive Value Propositions: Evidence from the Sports Industry. *International Journal of Online Marketing, 6*(2), 1–17. doi:10.4018/IJOM.2016040101

Shamsuzzoha, A. (2016). Management of Risk and Resilience within Collaborative Business Network. In R. Addo-Tenkorang, J. Kantola, P. Helo, & A. Shamsuzzoha (Eds.), *Supply Chain Strategies and the Engineer-to-Order Approach* (pp. 143–159). Hershey, PA: IGI Global. doi:10.4018/978-1-5225-0021-6.ch008

Shaqrah, A. A. (2018). Analyzing Business Intelligence Systems Based on 7s Model of McKinsey. *International Journal of Business Intelligence Research, 9*(1), 53–63. doi:10.4018/IJBIR.2018010104

Sharma, A. J. (2017). Enhancing Sustainability through Experiential Learning in Management Education. In N. Baporikar (Ed.), *Management Education for Global Leadership* (pp. 256–274). Hershey, PA: IGI Global. doi:10.4018/978-1-5225-1013-0.ch013

Shetty, K. P. (2017). Responsible Global Leadership: Ethical Challenges in Management Education. In N. Baporikar (Ed.), *Innovation and Shifting Perspectives in Management Education* (pp. 194–223). Hershey, PA: IGI Global. doi:10.4018/978-1-5225-1019-2.ch009

Sinthupundaja, J., & Kohda, Y. (2017). Effects of Corporate Social Responsibility and Creating Shared Value on Sustainability. *International Journal of Sustainable Entrepreneurship and Corporate Social Responsibility, 2*(1), 27–38. doi:10.4018/IJSECSR.2017010103

Škarica, I., & Hrgović, A. V. (2018). Implementation of Total Quality Management Principles in Public Health Institutes in the Republic of Croatia. *International Journal of Productivity Management and Assessment Technologies*, *6*(1), 1–16. doi:10.4018/IJPMAT.2018010101

Smuts, H., Kotzé, P., Van der Merwe, A., & Loock, M. (2017). Framework for Managing Shared Knowledge in an Information Systems Outsourcing Context. *International Journal of Knowledge Management*, *13*(4), 1–30. doi:10.4018/IJKM.2017100101

Soares, E. R., & Zaidan, F. H. (2016). Information Architecture and Business Modeling in Modern Organizations of Information Technology: Professional Career Plan in Organizations IT. In G. Jamil, J. Poças Rascão, F. Ribeiro, & A. Malheiro da Silva (Eds.), *Handbook of Research on Information Architecture and Management in Modern Organizations* (pp. 439–457). Hershey, PA: IGI Global. doi:10.4018/978-1-4666-8637-3.ch020

Sousa, M. J., Cruz, R., Dias, I., & Caracol, C. (2017). Information Management Systems in the Supply Chain. In G. Jamil, A. Soares, & C. Pessoa (Eds.), *Handbook of Research on Information Management for Effective Logistics and Supply Chains* (pp. 469–485). Hershey, PA: IGI Global. doi:10.4018/978-1-5225-0973-8.ch025

Spremic, M., Turulja, L., & Bajgoric, N. (2018). Two Approaches in Assessing Business Continuity Management Attitudes in the Organizational Context. In N. Bajgoric (Ed.), *Always-On Enterprise Information Systems for Modern Organizations* (pp. 159–183). Hershey, PA: IGI Global. doi:10.4018/978-1-5225-3704-5.ch008

Steenkamp, A. L. (2018). Some Insights in Computer Science and Information Technology. In *Examining the Changing Role of Supervision in Doctoral Research Projects: Emerging Research and Opportunities* (pp. 113–133). Hershey, PA: IGI Global. doi:10.4018/978-1-5225-2610-0.ch005

Studdard, N., Dawson, M., Burton, S. L., Jackson, N., Leonard, B., Quisenberry, W., & Rahim, E. (2016). Nurturing Social Entrepreneurship and Building Social Entrepreneurial Self-Efficacy: Focusing on Primary and Secondary Schooling to Develop Future Social Entrepreneurs. In Z. Fields (Ed.), *Incorporating Business Models and Strategies into Social Entrepreneurship* (pp. 154–175). Hershey, PA: IGI Global. doi:10.4018/978-1-4666-8748-6.ch010

Sun, Z. (2016). A Framework for Developing Management Intelligent Systems. *International Journal of Systems and Service-Oriented Engineering*, *6*(1), 37–53. doi:10.4018/IJSSOE.2016010103

Swami, B., & Mphele, G. T. (2016). Problems Preventing Growth of Small Entrepreneurs: A Case Study of a Few Small Entrepreneurs in Botswana Sub-Urban Areas. In N. Baporikar (Ed.), *Handbook of Research on Entrepreneurship in the Contemporary Knowledge-Based Global Economy* (pp. 479–508). Hershey, PA: IGI Global. doi:10.4018/978-1-4666-8798-1.ch020

Tabach, A., & Croteau, A. (2017). Configurations of Information Technology Governance Practices and Business Unit Performance. *International Journal of IT/Business Alignment and Governance*, *8*(2), 1–27. doi:10.4018/IJITBAG.2017070101

Talaue, G. M., & Iqbal, T. (2017). Assessment of e-Business Mode of Selected Private Universities in the Philippines and Pakistan. *International Journal of Online Marketing*, *7*(4), 63–77. doi:10.4018/IJOM.2017100105

Tam, G. C. (2017). Project Manager Sustainability Competence. In *Managerial Strategies and Green Solutions for Project Sustainability* (pp. 178–207). Hershey, PA: IGI Global. doi:10.4018/978-1-5225-2371-0.ch008

Tambo, T. (2018). Fashion Retail Innovation: About Context, Antecedents, and Outcome in Technological Change Projects. In I. Management Association (Ed.), Fashion and Textiles: Breakthroughs in Research and Practice (pp. 233-260). Hershey, PA: IGI Global. doi:10.4018/978-1-5225-3432-7.ch010

Tambo, T., & Mikkelsen, O. E. (2016). Fashion Supply Chain Optimization: Linking Make-to-Order Purchasing and B2B E-Commerce. In S. Joshi & R. Joshi (Eds.), *Designing and Implementing Global Supply Chain Management* (pp. 1–21). Hershey, PA: IGI Global. doi:10.4018/978-1-4666-9720-1.ch001

Tandon, K. (2016). Innovative Andragogy: The Paradigm Shift to Heutagogy. In S. Tiwari & L. Nafees (Eds.), *Innovative Management Education Pedagogies for Preparing Next-Generation Leaders* (pp. 238–257). Hershey, PA: IGI Global. doi:10.4018/978-1-4666-9691-4.ch014

Tantau, A. D., & Frăţilă, L. C. (2018). Information and Management System for Renewable Energy Business. In *Entrepreneurship and Business Development in the Renewable Energy Sector* (pp. 200–244). Hershey, PA: IGI Global. doi:10.4018/978-1-5225-3625-3.ch006

Teixeira, N., Pardal, P. N., & Rafael, B. G. (2018). Internationalization, Financial Performance, and Organizational Challenges: A Success Case in Portugal. In L. Carvalho (Ed.), *Handbook of Research on Entrepreneurial Ecosystems and Social Dynamics in a Globalized World* (pp. 379–423). Hershey, PA: IGI Global. doi:10.4018/978-1-5225-3525-6.ch017

Trad, A., & Kalpić, D. (2016). The E-Business Transformation Framework for E-Commerce Architecture-Modeling Projects. In I. Lee (Ed.), *Encyclopedia of E-Commerce Development, Implementation, and Management* (pp. 733–753). Hershey, PA: IGI Global. doi:10.4018/978-1-4666-9787-4.ch052

Trad, A., & Kalpić, D. (2016). The E-Business Transformation Framework for E-Commerce Control and Monitoring Pattern. In I. Lee (Ed.), *Encyclopedia of E-Commerce Development, Implementation, and Management* (pp. 754–777). Hershey, PA: IGI Global. doi:10.4018/978-1-4666-9787-4.ch053

Trad, A., & Kalpić, D. (2018). The Business Transformation Framework, Agile Project and Change Management. In M. Khosrow-Pour, D.B.A. (Ed.), Encyclopedia of Information Science and Technology, Fourth Edition (pp. 620-635). Hershey, PA: IGI Global. doi:10.4018/978-1-5225-2255-3.ch054

Trad, A., & Kalpić, D. (2018). The Business Transformation and Enterprise Architecture Framework: The Financial Engineering E-Risk Management and E-Law Integration. In B. Sergi, F. Fidanoski, M. Ziolo, & V. Naumovski (Eds.), *Regaining Global Stability After the Financial Crisis* (pp. 46–65). Hershey, PA: IGI Global. doi:10.4018/978-1-5225-4026-7.ch003

Turulja, L., & Bajgoric, N. (2018). Business Continuity and Information Systems: A Systematic Literature Review. In N. Bajgoric (Ed.), *Always-On Enterprise Information Systems for Modern Organizations* (pp. 60–87). Hershey, PA: IGI Global. doi:10.4018/978-1-5225-3704-5.ch004

van Wessel, R. M., de Vries, H. J., & Ribbers, P. M. (2016). Business Benefits through Company IT Standardization. In K. Jakobs (Ed.), *Effective Standardization Management in Corporate Settings* (pp. 34–53). Hershey, PA: IGI Global. doi:10.4018/978-1-4666-9737-9.ch003

Vargas-Hernández, J. G. (2017). Professional Integrity in Business Management Education. In N. Baporikar (Ed.), *Management Education for Global Leadership* (pp. 70–89). Hershey, PA: IGI Global. doi:10.4018/978-1-5225-1013-0.ch004

Vasista, T. G., & AlAbdullatif, A. M. (2017). Role of Electronic Customer Relationship Management in Demand Chain Management: A Predictive Analytic Approach. *International Journal of Information Systems and Supply Chain Management, 10*(1), 53–67. doi:10.4018/IJISSCM.2017010104

Vergidis, K. (2016). Rediscovering Business Processes: Definitions, Patterns, and Modelling Approaches. In P. Papajorgji, F. Pinet, A. Guimarães, & J. Papathanasiou (Eds.), *Automated Enterprise Systems for Maximizing Business Performance* (pp. 97–122). Hershey, PA: IGI Global. doi:10.4018/978-1-4666-8841-4.ch007

Vieru, D., & Bourdeau, S. (2017). Survival in the Digital Era: A Digital Competence-Based Multi-Case Study in the Canadian SME Clothing Industry. *International Journal of Social and Organizational Dynamics in IT, 6*(1), 17–34. doi:10.4018/IJSODIT.2017010102

Vijayan, G., & Kamarulzaman, N. H. (2017). An Introduction to Sustainable Supply Chain Management and Business Implications. In M. Khan, M. Hussain, & M. Ajmal (Eds.), *Green Supply Chain Management for Sustainable Business Practice* (pp. 27–50). Hershey, PA: IGI Global. doi:10.4018/978-1-5225-0635-5.ch002

Vlachvei, A., & Notta, O. (2017). Firm Competitiveness: Theories, Evidence, and Measurement. In A. Vlachvei, O. Notta, K. Karantininis, & N. Tsounis (Eds.), *Factors Affecting Firm Competitiveness and Performance in the Modern Business World* (pp. 1–42). Hershey, PA: IGI Global. doi:10.4018/978-1-5225-0843-4.ch001

von Rosing, M., Fullington, N., & Walker, J. (2016). Using the Business Ontology and Enterprise Standards to Transform Three Leading Organizations. *International Journal of Conceptual Structures and Smart Applications, 4*(1), 71–99. doi:10.4018/IJCSSA.2016010104

von Rosing, M., & von Scheel, H. (2016). Using the Business Ontology to Develop Enterprise Standards. *International Journal of Conceptual Structures and Smart Applications, 4*(1), 48–70. doi:10.4018/IJCSSA.2016010103

Walczak, S. (2016). Artificial Neural Networks and other AI Applications for Business Management Decision Support. *International Journal of Sociotechnology and Knowledge Development, 8*(4), 1–20. doi:10.4018/IJSKD.2016100101

Wamba, S. F., Akter, S., Kang, H., Bhattacharya, M., & Upal, M. (2016). The Primer of Social Media Analytics. *Journal of Organizational and End User Computing*, *28*(2), 1–12. doi:10.4018/JOEUC.2016040101

Wang, C., Schofield, M., Li, X., & Ou, X. (2017). Do Chinese Students in Public and Private Higher Education Institutes Perform at Different Level in One of the Leadership Skills: Critical Thinking?: An Exploratory Comparison. In V. Wang (Ed.), *Encyclopedia of Strategic Leadership and Management* (pp. 160–181). Hershey, PA: IGI Global. doi:10.4018/978-1-5225-1049-9.ch013

Wang, F., Raisinghani, M. S., Mora, M., & Wang, X. (2016). Strategic E-Business Management through a Balanced Scored Card Approach. In I. Lee (Ed.), *Encyclopedia of E-Commerce Development, Implementation, and Management* (pp. 361–386). Hershey, PA: IGI Global. doi:10.4018/978-1-4666-9787-4.ch027

Wang, J. (2017). Multi-Agent based Production Management Decision System Modelling for the Textile Enterprise. *Journal of Global Information Management*, *25*(4), 1–15. doi:10.4018/JGIM.2017100101

Wiedemann, A., & Gewald, H. (2017). Examining Cross-Domain Alignment: The Correlation of Business Strategy, IT Management, and IT Business Value. *International Journal of IT/Business Alignment and Governance*, *8*(1), 17–31. doi:10.4018/IJITBAG.2017010102

Wolf, R., & Thiel, M. (2018). Advancing Global Business Ethics in China: Reducing Poverty Through Human and Social Welfare. In S. Hipsher (Ed.), *Examining the Private Sector's Role in Wealth Creation and Poverty Reduction* (pp. 67–84). Hershey, PA: IGI Global. doi:10.4018/978-1-5225-3117-3.ch004

Wu, J., Ding, F., Xu, M., Mo, Z., & Jin, A. (2016). Investigating the Determinants of Decision-Making on Adoption of Public Cloud Computing in E-government. *Journal of Global Information Management*, *24*(3), 71–89. doi:10.4018/JGIM.2016070104

Xu, L., & de Vrieze, P. (2016). Building Situational Applications for Virtual Enterprises. In I. Lee (Ed.), *Encyclopedia of E-Commerce Development, Implementation, and Management* (pp. 715–724). Hershey, PA: IGI Global. doi:10.4018/978-1-4666-9787-4.ch050

Yablonsky, S. (2018). Innovation Platforms: Data and Analytics Platforms. In *Multi-Sided Platforms (MSPs) and Sharing Strategies in the Digital Economy: Emerging Research and Opportunities* (pp. 72–95). Hershey, PA: IGI Global. doi:10.4018/978-1-5225-5457-8.ch003

Yusoff, A., Ahmad, N. H., & Halim, H. A. (2017). Agropreneurship among Gen Y in Malaysia: The Role of Academic Institutions. In N. Ahmad, T. Ramayah, H. Halim, & S. Rahman (Eds.), *Handbook of Research on Small and Medium Enterprises in Developing Countries* (pp. 23–47). Hershey, PA: IGI Global. doi:10.4018/978-1-5225-2165-5.ch002

Zanin, F., Comuzzi, E., & Costantini, A. (2018). The Effect of Business Strategy and Stock Market Listing on the Use of Risk Assessment Tools. In *Management Control Systems in Complex Settings: Emerging Research and Opportunities* (pp. 145–168). Hershey, PA: IGI Global. doi:10.4018/978-1-5225-3987-2.ch007

Zgheib, P. W. (2017). Corporate Innovation and Intrapreneurship in the Middle East. In P. Zgheib (Ed.), *Entrepreneurship and Business Innovation in the Middle East* (pp. 37–56). Hershey, PA: IGI Global. doi:10.4018/978-1-5225-2066-5.ch003

Index

N

O

P

Q

R

S

T

U

www.ingramcontent.com/pod-product-compliance
Lightning Source LLC
LaVergne TN
LVHW061534290926
849162LV00004B/788
* 9 7 8 1 5 2 2 5 6 9 7 4 9 *